Introduction to
Semi-Supervised Learning

Synthesis Lectures on Artificial Intelligence and Machine Learning

Editors
Ronald J. Brachman, *Yahoo! Research*
Thomas Dietterich, *Oregon State University*

Introduction to Semi-Supervised Learning
Xiaojin Zhu and Andrew B. Goldberg

ISBN: 978-3-031-00420-9   paperback
ISBN: 978-3-031-01548-9   ebook

DOI 10.1007/978-3-031-01548-9

A Publication in the Springer series
SYNTHESIS LECTURES ON ARTIFICIAL INTELLIGENCE AND MACHINE LEARNING

Lecture #6
Series Editors: Ronald J. Brachman, *Yahoo! Research*
                Thomas Dietterich, *Oregon State University*

Series ISSN
Synthesis Lectures on Artificial Intelligence and Machine Learning
Print 1939-4608   Electronic 1939-4616

Introduction to
Semi-Supervised Learning

Xiaojin Zhu and Andrew B. Goldberg

University of Wisconsin, Madison

SYNTHESIS LECTURES ON ARTIFICIAL INTELLIGENCE AND MACHINE LEARNING #6

ABSTRACT

Semi-supervised learning is a learning paradigm concerned with the study of how computers and natural systems such as humans learn in the presence of both labeled and unlabeled data. Traditionally, learning has been studied either in the unsupervised paradigm (e.g., clustering, outlier detection) where all the data is unlabeled, or in the supervised paradigm (e.g., classification, regression) where all the data is labeled. The goal of semi-supervised learning is to understand how combining labeled and unlabeled data may change the learning behavior, and design algorithms that take advantage of such a combination. Semi-supervised learning is of great interest in machine learning and data mining because it can use readily available unlabeled data to improve supervised learning tasks when the labeled data is scarce or expensive. Semi-supervised learning also shows potential as a quantitative tool to understand human category learning, where most of the input is self-evidently unlabeled. In this introductory book, we present some popular semi-supervised learning models, including self-training, mixture models, co-training and multiview learning, graph-based methods, and semi-supervised support vector machines. For each model, we discuss its basic mathematical formulation. The success of semi-supervised learning depends critically on some underlying assumptions. We emphasize the assumptions made by each model and give counterexamples when appropriate to demonstrate the limitations of the different models. In addition, we discuss semi-supervised learning for cognitive psychology. Finally, we give a computational learning theoretic perspective on semi-supervised learning, and we conclude the book with a brief discussion of open questions in the field.

KEYWORDS

semi-supervised learning, transductive learning, self-training, Gaussian mixture model, expectation maximization (EM), cluster-then-label, co-training, multiview learning, mincut, harmonic function, label propagation, manifold regularization, semi-supervised support vector machines (S3VM), transductive support vector machines (TSVM), entropy regularization, human semi-supervised learning

To our parents

Yu and Jingquan
Susan and Steven Goldberg

with much love and gratitude.

Contents

Preface

The book is a beginner's guide to semi-supervised learning. It is aimed at advanced undergraduates, entry-level graduate students and researchers in areas as diverse as Computer Science, Electrical Engineering, Statistics, and Psychology. The book assumes that the reader is familiar with elementary calculus, probability and linear algebra. It is helpful, but not necessary, for the reader to be familiar with statistical machine learning, as we will explain the essential concepts in order for this book to be self-contained. Sections containing more advanced materials are marked with a star. We also provide a basic mathematical reference in Appendix A.

Our focus is on semi-supervised model assumptions and computational techniques. We intentionally avoid competition-style benchmark evaluations. This is because, in general, semi-supervised learning models are sensitive to various settings, and no benchmark that we know of can characterize the full potential of a given model on all tasks. Instead, we will often use simple artificial problems to "break" the models in order to reveal their assumptions. Such analysis is not frequently encountered in the literature.

Semi-supervised learning has grown into a large research area within machine learning. For example, a search for the phrase "semi-supervised" in May 2009 yielded more than 8000 papers in Google Scholar. While we attempt to provide a basic coverage of semi-supervised learning, the selected topics are not able to reflect the most recent advances in the field. We provide a "bibliographical notes" section at the end of each chapter for the reader to dive deeper into the topics.

We would like to express our sincere thanks to Thorsten Joachims and the other reviewers for their constructive reviews that greatly improved the book. We thank Robert Nowak for his excellent learning theory lecture notes, from which we take some materials for Section 8.1. Our thanks also go to Bryan Gibson, Tushar Khot, Robert Nosofsky, Timothy Rogers, and Zhiting Xu for their valuable comments.

We hope you enjoy the book.

Xiaojin Zhu and Andrew B. Goldberg
Madison, Wisconsin

C H A P T E R   1

Introduction to Statistical Machine Learning

We start with a gentle introduction to statistical machine learning. Readers familiar with machine learning may wish to skip directly to Section 2, where we introduce semi-supervised learning.

Example 1.1. You arrive at an extrasolar planet and are welcomed by its resident little green men. You observe the weight and height of 100 little green men around you, and plot the measurements in Figure 1.1. What can you learn from this data?

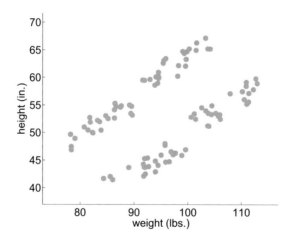

Figure 1.1: The weight and height of 100 little green men from the extrasolar planet. Each green dot is an instance, represented by two features: weight and height.

This is a typical example of a machine learning scenario (except the little green men part). We can perform several tasks using this data: group the little green men into subcommunities based on weight and/or height, identify individuals with extreme (possibly erroneous) weight or height values, try to predict one measurement based on the other, etc. Before exploring such machine learning tasks, let us begin with some definitions.

1.1 THE DATA

Definition 1.2. Instance. An *instance* $\mathbf{x}$ represents a specific object. The instance is often represented by a D-dimensional *feature vector* $\mathbf{x} = (x_1, \ldots, x_D) \in \mathbb{R}^D$, where each dimension is called a *feature*. The length D of the feature vector is known as the dimensionality of the feature vector.

The feature representation is an abstraction of the objects. It essentially ignores all other information not represented by the features. For example, two little green men with the same weight and height, but with different names, will be regarded as indistinguishable by our feature representation. Note we use boldface $\mathbf{x}$ to denote the whole instance, and x_d to denote the d-th feature of $\mathbf{x}$. In our example, an instance is a specific little green man; the feature vector consists of $D = 2$ features: x_1 is the weight, and x_2 is the height. Features can also take discrete values. When there are multiple instances, we will use x_{id} to denote the i-th instance's d-th feature.

Definition 1.3. Training Sample. A *training sample* is a collection of instances $\{\mathbf{x}_i\}_{i=1}^{n} = \{\mathbf{x}_1, \ldots, \mathbf{x}_n\}$, which acts as the input to the learning process. We assume these instances are sampled independently from an underlying distribution $P(\mathbf{x})$, which is unknown to us. We denote this by $\{\mathbf{x}_i\}_{i=1}^{n} \overset{\text{i.i.d.}}{\sim} P(\mathbf{x})$, where i.i.d. stands for independent and identically distributed.

In our example, the training sample consists of $n = 100$ instances $\mathbf{x}_1, \ldots, \mathbf{x}_{100}$. A training sample is the "experience" given to a learning algorithm. What the algorithm can learn from it, however, varies. In this chapter, we introduce two basic learning paradigms: *unsupervised learning* and *supervised learning*.

1.2 UNSUPERVISED LEARNING

Definition 1.4. Unsupervised learning. Unsupervised learning algorithms work on a training sample with n instances $\{\mathbf{x}_i\}_{i=1}^{n}$. There is no teacher providing supervision as to how individual instances should be handled—this is the defining property of unsupervised learning. Common unsupervised learning tasks include:

- clustering, where the goal is to separate the n instances into groups;

- novelty detection, which identifies the few instances that are very different from the majority;

- dimensionality reduction, which aims to represent each instance with a lower dimensional feature vector while maintaining key characteristics of the training sample.

Among the unsupervised learning tasks, the one most relevant to this book is *clustering*, which we discuss in more detail.

Definition 1.5. Clustering. Clustering splits $\{\mathbf{x}_i\}_{i=1}^{n}$ into k clusters, such that instances in the same cluster are similar, and instances in different clusters are dissimilar. The number of clusters k may be specified by the user, or may be inferred from the training sample itself.

How many clusters do you find in the little green men data in Figure 1.1? Perhaps $k = 2$, $k = 4$, or more. Without further assumptions, either one is acceptable. Unlike in supervised learning (introduced in the next section), there is no teacher that tells us which instances should be in each cluster.

There are many clustering algorithms. We introduce a particularly simple one, *hierarchical agglomerative clustering*, to make unsupervised learning concrete.

Algorithm 1.6. Hierarchical Agglomerative Clustering.

Input: a training sample $\{\mathbf{x}_i\}_{i=1}^n$; a distance function $d()$.
1. Initially, place each instance in its own cluster (called a singleton cluster).
*2. **while** (number of clusters > 1) **do**:*
3.   Find the closest cluster pair A, B, i.e., they minimize $d(A, B)$.
4.   Merge A, B to form a new cluster.
***Output**: a binary tree showing how clusters are gradually merged from singletons to a root cluster, which contains the whole training sample.*

This clustering algorithm is simple. The only thing unspecified is the distance function $d()$. If $\mathbf{x}_i, \mathbf{x}_j$ are two singleton clusters, one way to define $d(\mathbf{x}_i, \mathbf{x}_j)$ is the Euclidean distance between them:

$$d(\mathbf{x}_i, \mathbf{x}_j) = \|\mathbf{x}_i - \mathbf{x}_j\| = \sqrt{\sum_{s=1}^{D}(x_{is} - x_{js})^2}. \tag{1.1}$$

We also need to define the distance between two non-singleton clusters A, B. There are multiple possibilities: one can define it to be the distance between the closest pair of points in A and B, the distance between the farthest pair, or some average distance. For simplicity, we will use the first option, also known as *single linkage*:

$$d(A, B) = \min_{\mathbf{x}\in A, \mathbf{x}'\in B} d(\mathbf{x}, \mathbf{x}'). \tag{1.2}$$

It is not necessary to fully grow the tree until only one cluster remains: the clustering algorithm can be stopped at any point if $d()$ exceeds some threshold, or the number of clusters reaches a predetermined number k.

Figure 1.2 illustrates the results of hierarchical agglomerative clustering for $k = 2, 3, 4$, respectively. The clusters certainly look fine. But because there is no information on how each instance *should* be clustered, it can be difficult to objectively evaluate the result of clustering algorithms.

1.3   SUPERVISED LEARNING

Suppose you realize that your alien hosts have a gender: female or male (so they should not all be called little green men after all). You may now be interested in predicting the gender of a particular

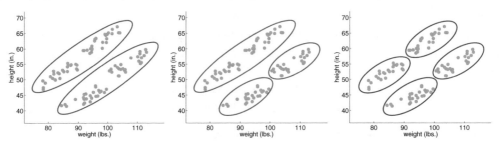

Figure 1.2: Hierarchical agglomerative clustering results for $k = 2, 3, 4$ on the 100 little green men data.

alien from his or her weight and height. Alternatively, you may want to predict whether an alien is a juvenile or an adult using weight and height. To explain how to approach these tasks, we need more definitions.

Definition 1.7. Label.   A label y is the desired prediction on an instance $\mathbf{x}$.

Labels may come from a finite set of values, e.g., {female, male}. These distinct values are called *classes*. The classes are usually encoded by integer numbers, e.g., female $= -1$, male $= 1$, and thus $y \in \{-1, 1\}$. This particular encoding is often used for binary (two-class) labels, and the two classes are generically called the negative class and the positive class, respectively. For problems with more than two classes, a traditional encoding is $y \in \{1, \ldots, C\}$, where C is the number of classes. In general, such encoding does not imply structure in the classes. That is to say, the two classes encoded by $y = 1$ and $y = 2$ are not necessarily closer than the two classes $y = 1$ and $y = 3$. Labels may also take continuous values in $\mathbb{R}$. For example, one may attempt to predict the blood pressure of little green aliens based on their height and weight.

In supervised learning, the training sample consists of pairs, each containing an instance $\mathbf{x}$ and a label y: $\{(\mathbf{x}_i, y_i)\}_{i=1}^{n}$. One can think of y as the label on $\mathbf{x}$ provided by a teacher, hence the name *supervised* learning. Such (instance, label) pairs are called *labeled data*, while instances alone without labels (as in unsupervised learning) are called *unlabeled data*. We are now ready to define supervised learning.

Definition 1.8. Supervised learning.   Let the domain of instances be $\mathcal{X}$, and the domain of labels be $\mathcal{Y}$. Let $P(\mathbf{x}, y)$ be an (unknown) joint probability distribution on instances and labels $\mathcal{X} \times \mathcal{Y}$. Given a training sample $\{(\mathbf{x}_i, y_i)\}_{i=1}^{n} \overset{\text{i.i.d.}}{\sim} P(\mathbf{x}, y)$, supervised learning trains a function $f : \mathcal{X} \mapsto \mathcal{Y}$ in some function family $\mathcal{F}$, with the goal that $f(\mathbf{x})$ predicts the true label y on future data $\mathbf{x}$, where $(\mathbf{x}, y) \overset{\text{i.i.d.}}{\sim} P(\mathbf{x}, y)$ as well.

Depending on the domain of label y, supervised learning problems are further divided into *classification* and *regression*:

Definition 1.9. Classification.   Classification is the supervised learning problem with discrete classes $\mathcal{Y}$. The function f is called a *classifier*.

Definition 1.10. Regression.   Regression is the supervised learning problem with continuous $\mathcal{Y}$. The function f is called a *regression function*.

What exactly is a good f? The best f is by definition

$$f^* = \operatorname*{argmin}_{f \in \mathcal{F}} \mathbb{E}_{(\mathbf{x},y)\sim P}\left[c(\mathbf{x}, y, f(\mathbf{x}))\right], \tag{1.3}$$

where argmin means "finding the f that minimizes the following quantity". $\mathbb{E}_{(\mathbf{x},y)\sim P}[\cdot]$ is the expectation over random test data drawn from P. Readers not familiar with this notation may wish to consult Appendix A. $c(\cdot)$ is a *loss function* that determines the cost or impact of making a prediction $f(\mathbf{x})$ that is different from the true label y. Some typical loss functions will be discussed shortly. Note we limit our attention to some function family $\mathcal{F}$, mostly for computational reasons. If we remove this limitation and consider all possible functions, the resulting f^* is the *Bayes optimal predictor*, the best one can hope for on average. For the distribution P, this function will incur the lowest possible loss when making predictions. The quantity $\mathbb{E}_{(\mathbf{x},y)\sim P}[c(\mathbf{x}, y, f^*(\mathbf{x}))]$ is known as the *Bayes error*. However, the Bayes optimal predictor may not be in $\mathcal{F}$ in general. Our goal is to find the $f \in \mathcal{F}$ that is as close to the Bayes optimal predictor as possible.

It is worth noting that the underlying distribution $P(\mathbf{x}, y)$ is unknown to us. Therefore, it is not possible to directly find f^*, or even to measure any predictor f's performance, for that matter. Here lies the fundamental difficulty of statistical machine learning: one has to *generalize* the prediction from a finite training sample to any unseen test data. This is known as *induction*.

To proceed, a seemingly reasonable approximation is to gauge f's performance using training sample error. That is, to replace the unknown expectation by the average over the training sample:

Definition 1.11. Training sample error.   Given a training sample $\{(\mathbf{x}_i, y_i)\}_{i=1}^{n}$, the training sample error is

$$\frac{1}{n}\sum_{i=1}^{n} c(\mathbf{x}_i, y_i, f(\mathbf{x}_i)). \tag{1.4}$$

For classification, one commonly used loss function is the 0-1 loss $c(\mathbf{x}, y, f(\mathbf{x})) \equiv (f(\mathbf{x}_i) \neq y_i)$:

$$\frac{1}{n}\sum_{i=1}^{n}(f(\mathbf{x}_i) \neq y_i), \tag{1.5}$$

where $f(\mathbf{x}) \neq y$ is 1 if f predicts a different class than y on x, and 0 otherwise. For regression, one commonly used loss function is the squared loss $c(\mathbf{x}, y, f(\mathbf{x})) \equiv (f(\mathbf{x}_i) - y_i)^2$:

$$\frac{1}{n} \sum_{i=1}^{n} (f(\mathbf{x}_i) - y_i)^2. \tag{1.6}$$

Other loss functions will be discussed as we encounter them later in the book.

It might be tempting to seek the f that minimizes training sample error. However, this strategy is flawed: such an f will tend to *overfit* the particular training sample. That is, it will likely fit itself to the statistical noise in the particular training sample. It will learn more than just the true relationship between $\mathcal{X}$ and $\mathcal{Y}$. Such an overfitted predictor will have small training sample error, but is likely to perform less well on future test data. A sub-area within machine learning called computational learning theory studies the issue of overfitting. It establishes rigorous connections between the training sample error and the true error, using a formal notion of complexity such as the Vapnik-Chervonenkis dimension or Rademacher complexity. We provide a concise discussion in Section 8.1. Informed by computational learning theory, one reasonable training strategy is to seek an f that "almost" minimizes the training sample error, while *regularizing* f so that it is not too complex in a certain sense. Interested readers can find the references in the bibliographical notes.

To estimate f's future performance, one can use a separate sample of labeled instances, called the *test sample*: $\{(\mathbf{x}_j, y_j)\}_{j=n+1}^{n+m} \overset{\text{i.i.d.}}{\sim} P(\mathbf{x}, y)$. A test sample is not used during training, and therefore provides a faithful (unbiased) estimation of future performance.

Definition 1.12. Test sample error.   The corresponding test sample error for classification with 0-1 loss is

$$\frac{1}{m} \sum_{j=n+1}^{n+m} (f(\mathbf{x}_j) \neq y_j), \tag{1.7}$$

and for regression with squared loss is

$$\frac{1}{m} \sum_{j=n+1}^{n+m} (f(\mathbf{x}_j) - y_j)^2. \tag{1.8}$$

In the remainder of the book, we focus on classification due to its prevalence in semi-supervised learning research. Most ideas discussed also apply to regression, though.

As a concrete example of a supervised learning method, we now introduce a simple classification algorithm: k-nearest-neighbor (kNN).

Algorithm 1.13. k-nearest-neighbor classifier.

Input: Training data $(\mathbf{x}_1, y_1), \ldots, (\mathbf{x}_n, y_n)$; *distance function* $d()$;
    number of neighbors k; test instance $\mathbf{x}^*$

1. Find the k training instances $\mathbf{x}_{i_1}, \ldots, \mathbf{x}_{i_k}$ closest to $\mathbf{x}^*$ under distance $d()$.
2. Output y^* as the majority class of $y_{i_1}, \ldots, y_{i_k}$. Break ties randomly.

Being a D-dimensional feature vector, the test instance $\mathbf{x}^*$ can be viewed as a point in D-dimensional feature space. A classifier assigns a label to each point in the feature space. This divides the feature space into decision regions within which points have the same label. The boundary separating these regions is called the *decision boundary* induced by the classifier.

Example 1.14.   Consider two classification tasks involving the little green aliens. In the first task in Figure 1.3(a), the task is gender classification from weight and height. The symbols are training data. Each training instance has a label: female (red cross) or male (blue circle). The decision regions from a 1NN classifier are shown as white and gray. In the second task in Figure 1.3(b), the task is age classification on the same sample of training instances. The training instances now have different labels: juvenile (red cross) or adult (blue circle). Again, the decision regions of 1NN are shown. Notice that, for the same training instances but different classification goals, the decision boundary can be quite different. Naturally, this is a property unique to supervised learning, since unsupervised learning does not use any particular set of labels at all.

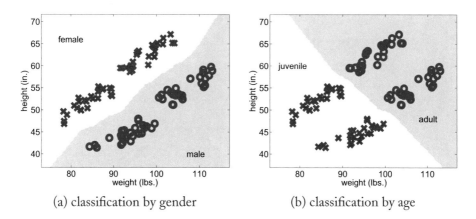

(a) classification by gender   (b) classification by age

Figure 1.3: Classify by gender or age from a training sample of 100 little green aliens, with 1-nearest-neighbor decision regions shown.

In this chapter, we introduced statistical machine learning as a foundation for the rest of the book. We presented the unsupervised and supervised learning settings, along with concrete examples of each. In the next chapter, we provide an overview of semi-supervised learning, which falls somewhere between these two. Each subsequent chapter will present specific families of semi-supervised learning algorithms.

BIBLIOGRAPHICAL NOTES

There are many excellent books written on statistical machine learning. For example, readers interested in the methodologies can consult the introductory textbook [131], and the comprehensive textbooks [19, 81]. For grounding of machine learning in classic statistics, see [184]. For computational learning theory, see [97, 176] for the Vapnik-Chervonenkis (VC) dimension and Probably Approximately Correct (PAC) learning framework, and Chapter 4 in [153] for an introduction to the Rademacher complexity. For a perspective from information theory, see [119]. For a perspective that views machine learning as an important part of artificial intelligence, see [147].

CHAPTER 2

Overview of Semi-Supervised Learning

2.1 LEARNING FROM BOTH LABELED AND UNLABELED DATA

As the name suggests, *semi-supervised learning* is somewhere between unsupervised and supervised learning. In fact, most semi-supervised learning strategies are based on extending either unsupervised or supervised learning to include additional information typical of the other learning paradigm. Specifically, semi-supervised learning encompasses several different settings, including:

- *Semi-supervised classification.* Also known as classification with labeled and unlabeled data (or partially labeled data), this is an extension to the supervised classification problem. The training data consists of both l labeled instances $\{(\mathbf{x}_i, y_i)\}_{i=1}^{l}$ and u unlabeled instances $\{\mathbf{x}_j\}_{j=l+1}^{l+u}$. One typically assumes that there is much more unlabeled data than labeled data, i.e., $u \gg l$. The goal of semi-supervised classification is to train a classifier f from both the labeled and unlabeled data, such that it is better than the supervised classifier trained on the labeled data alone.

- *Constrained clustering.* This is an extension to unsupervised clustering. The training data consists of unlabeled instances $\{x_i\}_{j=1}^{n}$, as well as some "supervised information" about the clusters. For example, such information can be so-called *must-link* constraints, that two instances $\mathbf{x}_i, \mathbf{x}_j$ must be in the same cluster; and *cannot-link* constraints, that $\mathbf{x}_i, \mathbf{x}_j$ cannot be in the same cluster. One can also constrain the size of the clusters. The goal of constrained clustering is to obtain better clustering than the clustering from unlabeled data alone.

There are other semi-supervised learning settings, including regression with labeled and unlabeled data, dimensionality reduction with labeled instances whose reduced feature representation is given, and so on. This book will focus on semi-supervised classification.

The study of semi-supervised learning is motivated by two factors: its practical value in building better computer algorithms, and its theoretical value in understanding learning in machines and humans.

Semi-supervised learning has tremendous practical value. In many tasks, there is a paucity of labeled data. The labels y may be difficult to obtain because they require human annotators, special devices, or expensive and slow experiments. For example,

- In speech recognition, an instance $\mathbf{x}$ is a speech utterance, and the label y is the corresponding transcript. For example, here are some detailed phonetic transcripts of words as they are spoken:

film $\Rightarrow$ `f ih_n uh_gl_n m`
be all $\Rightarrow$ `bcl b iy iy_tr ao_tr ao l_dl`

Accurate transcription by human expert annotators can be extremely time consuming: it took as long as 400 hours to transcribe 1 hour of speech at the phonetic level for the Switchboard telephone conversational speech data [71] (recordings of randomly paired participants discussing various topics such as social, economic, political, and environmental issues).

- In natural language parsing, an instance **x** is a sentence, and the label y is the corresponding parse tree. An example parse tree for the Chinese sentence "The National Track and Field Championship has finished." is shown below.

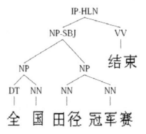

The training data, consisting of (sentence, parse tree) pairs, is known as a treebank. Treebanks are time consuming to construct, and require the expertise of linguists: For a mere 4000 sentences in the Penn Chinese Treebank, experts took two years to manually create the corresponding parse trees.

- In spam filtering, an instance **x** is an email, and the label y is the user's judgment (spam or ham). In this situation, the bottleneck is an average user's patience to label a large number of emails.

- In video surveillance, an instance **x** is a video frame, and the label y is the identity of the object in the video. Manually labeling the objects in a large number of surveillance video frames is tedious and time consuming.

- In protein 3D structure prediction, an instance **x** is a DNA sequence, and the label y is the 3D protein folding structure. It can take months of expensive laboratory work by expert crystallographers to identify the 3D structure of a single protein.

While labeled data $(\mathbf{x}, y)$ is difficult to obtain in these domains, unlabeled data **x** is available in large quantity and easy to collect: speech utterances can be recorded from radio broadcasts; text sentences can be crawled from the World Wide Web; emails are sitting on the mail server; surveillance cameras run 24 hours a day; and DNA sequences of proteins are readily available from gene databases. However, traditional supervised learning methods cannot use unlabeled data in training classifiers.

Semi-supervised learning is attractive because it can potentially utilize both labeled and unlabeled data to achieve better performance than supervised learning. From a different perspective, semi-supervised learning may achieve the same level of performance as supervised learning, but with fewer labeled instances. This reduces the annotation effort, which leads to reduced cost. We will present several computational models in Chapters 3,4,5, 6.

Semi-supervised learning also provides a computational model of how humans learn from labeled and unlabeled data. Consider the task of concept learning in children, which is similar to classification: an instance $\mathbf{x}$ is an object (e.g., an animal), and the label y is the corresponding concept (e.g., dog). Young children receive labeled data from teachers (e.g., Daddy points to a brown animal and says "dog!"). But more often they observe various animals by themselves without receiving explicit labels. It seems self-evident that children are able to combine labeled and unlabeled data to facilitate concept learning. The study of semi-supervised learning is therefore an opportunity to bridge machine learning and human learning. We will discuss some recent studies in Chapter 7.

2.2   HOW IS SEMI-SUPERVISED LEARNING POSSIBLE?

At first glance, it might seem paradoxical that one can learn anything about a predictor $f : \mathcal{X} \mapsto \mathcal{Y}$ from unlabeled data. After all, f is about the mapping from instance $\mathbf{x}$ to label y, yet unlabeled data does not provide any examples of such a mapping. The answer lies in the assumptions one makes about the link between the distribution of unlabeled data $P(\mathbf{x})$ and the target label.

Figure 2.1 shows a simple example of semi-supervised learning. Let each instance be represented by a one-dimensional feature $x \in \mathbb{R}$. There are two classes: positive and negative. Consider the following two scenarios:

1. In supervised learning, we are given only two labeled training instances $(\mathbf{x}_1, y_1) = (-1, -)$ and $(\mathbf{x}_2, y_2) = (1, +)$, shown as the red and blue symbols in the figure, respectively. The best estimate of the decision boundary is obviously $\mathbf{x} = 0$: all instances with $\mathbf{x} < 0$ should be classified as $y = -$, while those with $\mathbf{x} \geq 0$ as $y = +$.

2. In addition, we are also given a large number of unlabeled instances, shown as green dots in the figure. The correct class labels for these unlabeled examples are unknown. However, we observe that they form two groups. *Under the assumption* that instances in each class form a coherent group (e.g., $p(\mathbf{x}|y)$ is a Gaussian distribution, such that the instances from each class center around a central mean), this unlabeled data gives us more information. Specifically, it seems that the two labeled instances are not the most prototypical examples for the classes. Our *semi-supervised* estimate of the decision boundary should be between the two groups instead, at $\mathbf{x} \approx 0.4$.

If our assumption is true, then using both labeled and unlabeled data gives us a more reliable estimate of the decision boundary. Intuitively, the distribution of unlabeled data helps to identify regions with the same label, and the few labeled data then provide the actual labels. In this book, we will introduce a few other commonly used semi-supervised learning assumptions.

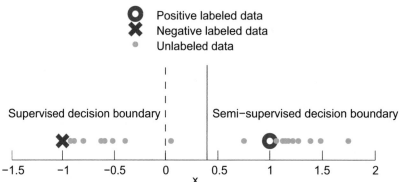

Figure 2.1: A simple example to demonstrate how semi-supervised learning is possible.

2.3   INDUCTIVE VS. TRANSDUCTIVE SEMI-SUPERVISED LEARNING

There are actually two slightly different semi-supervised learning settings, namely inductive and transductive semi-supervised learning. Recall that in supervised classification, the training sample is fully labeled, so one is always interested in the performance on future test data. In semi-supervised classification, however, the training sample contains some unlabeled data. Therefore, there are two distinct goals. One is to predict the labels on future test data. The other goal is to predict the labels on the unlabeled instances in the training sample. We call the former *inductive semi-supervised learning*, and the latter *transductive learning*.

Definition 2.1. Inductive semi-supervised learning.   Given a training sample $\{(\mathbf{x}_i, y_i)\}_{i=1}^{l}$, $\{\mathbf{x}_j\}_{j=l+1}^{l+u}$, inductive semi-supervised learning learns a function $f : \mathcal{X} \mapsto \mathcal{Y}$ so that f is expected to be a good predictor on future data, beyond $\{\mathbf{x}_j\}_{j=l+1}^{l+u}$.

Like in supervised learning, one can estimate the performance on future data by using a separate test sample $\{(\mathbf{x}_k, y_k)\}_{k=1}^{m}$, which is not available during training.

Definition 2.2. Transductive learning.   Given a training sample $\{(\mathbf{x}_i, y_i)\}_{i=1}^{l}$, $\{\mathbf{x}_j\}_{j=l+1}^{l+u}$, transductive learning trains a function $f : \mathcal{X}^{l+u} \mapsto \mathcal{Y}^{l+u}$ so that f is expected to be a good predictor on the unlabeled data $\{\mathbf{x}_j\}_{j=l+1}^{l+u}$. Note f is defined only on the given training sample, and is not required to make predictions outside. It is therefore a simpler function.

There is an interesting analogy: inductive semi-supervised learning is like an in-class exam, where the questions are not known in advance, and a student needs to prepare for all possible questions; in contrast, transductive learning is like a take-home exam, where the student knows the exam questions and needs not prepare beyond those.

2.4   CAVEATS

It seems reasonable that semi-supervised learning can use additional unlabeled data, which by it-self does not carry information on the mapping $\mathcal{X} \mapsto \mathcal{Y}$, to learn a better predictor f. As mentioned earlier, the key lies in the semi-supervised model *assumptions* about the link between the marginal distribution $P(\mathbf{x})$ and the conditional distribution $P(y|\mathbf{x})$. There are several different semi-supervised learning methods, and each makes slightly different assumptions about this link. These methods include self-training, probabilistic generative models, co-training, graph-based models, semi-supervised support vector machines, and so on. In the next several chapters, we will go through these models and discuss their assumptions. In Section 8.2, we will also give some theoretic justification. Empirically, these semi-supervised learning models do produce better classifiers than supervised learning on some data sets.

However, it is worth pointing out that blindly selecting a semi-supervised learning method for a specific task will not necessarily improve performance over supervised learning. In fact, unlabeled data can lead to *worse* performance with the wrong link assumptions. The following example demonstrates this sensitivity to model assumptions by comparing supervised learning performance with several semi-supervised learning approaches on a simple classification problem. Don't worry if these approaches appear mysterious; we will explain how they work in detail in the rest of the book. For now, the main point is that semi-supervised learning performance depends on the correctness of the assumptions made by the model in question.

Example 2.3.   Consider a classification task where there are two classes, each with a Gaussian distribution. The two Gaussian distributions heavily overlap (top panel of Figure 2.2). The true decision boundary lies in the middle of the two distributions, shown as a dotted line. Since we know the true distributions, we can compute test sample error rates based on the probability mass of each Gaussian that falls on the incorrect side of the decision boundary. Due to the overlapping class distributions, the optimal error rate (i.e., the Bayes error) is 21.2%.

For supervised learning, the learned decision boundary is in the middle of the two labeled instances, and the unlabeled instances are ignored. See, for example, the thick solid line in the second panel of Figure 2.2. We note that it is away from the true decision boundary, because the two labeled instances are randomly sampled. If we were to draw two other labeled instances, the learned decision boundary would change, but most likely would still be off (see other panels of Figure 2.2). On average, the *expected* learned decision boundary will coincide with the true boundary, but for any given draw of labeled data it will be off quite a bit. We say that the learned boundary has high variance. To evaluate supervised learning, and the semi-supervised learning methods introduced below, we drew 1000 training samples, each with one labeled and 99 unlabeled instances per class. In contrast to the optimal decision boundary, the decision boundaries found using supervised learning have an average test sample error rate of 31.6%. The average decision boundary lies at 0.02 (compared to the optimal boundary of 0), but has standard deviation of 0.72.

Now without presenting the details, we show the learned decision boundaries of three semi-supervised learning models on the training data. These models will be presented in detail in later

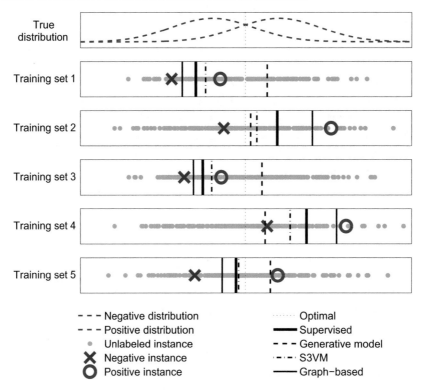

Figure 2.2: Two classes drawn from overlapping Gaussian distributions (top panel). Decision boundaries learned by several algorithms are shown for five random samples of labeled and unlabeled training samples.

chapters. The first one is a probabilistic generative model with two Gaussian distributions learned with EM (Chapter 3)—this model makes the correct model assumption. The decision boundaries are shown in Figure 2.2 as dashed lines. In this case, the boundaries tend to be closer to the true boundary and similar to one another, i.e., this algorithm has low variance. The 1000-trial average test sample error rate for this algorithm is 30.2%. The average decision boundary is at -0.003 with a standard deviation of 0.55, indicating the algorithm is both more accurate and more stable than the supervised model.

The second model is a semi-supervised support vector machine (Chapter 6), which assumes that the decision boundary should not pass through dense unlabeled data regions. However, since the two classes strongly overlap, the true decision boundary actually passes through the densest region. Therefore, the model assumption does not entirely match the task. The learned decision boundaries are shown in Figure 2.2 as dash-dotted lines.[1] The result is better than supervised classification and performs about the same as the probabilistic generative model that makes the correct model

---

[1]The semi-supervised support vector machine results were obtained using transductive SVM code similar to SVM-light.

assumption. The average test sample error rate here is 29.6%, with an average decision boundary of 0.01 (standard deviation 0.48). Despite the wrong model assumption, this approach uses knowledge that the two classes contain roughly the same number of instances, so the decision boundaries are drawn toward the center. This might explain the surprisingly good performance compared to the correct model.

The third approach is a graph-based model (Chapter 5), with a typical way to generate the graph: any two instances in the labeled and unlabeled data are connected by an edge. The edge weight is large if the two instances are close to each other, and small if they are far away. The model assumption is that instances connected with large-weight edges tend to have the same label. However, in this particular example where the two classes overlap, instances from different classes can be quite close and connected by large-weight edges. Therefore, the model assumption does not match the task either. The results using this model are shown in Figure 2.2 as thin solid lines.[2] The graph-based models' average test sample error rate is 36.4%, with an average decision boundary at 0.03 (standard deviation 1.23). The graph-based model is inappropriate for this task and performs even worse than supervised learning.

As the above example shows, the model assumption plays an important role in semi-supervised learning. It makes up for the lack of labeled data, and can determine the quality of the predictor. However, making the right assumptions (or detecting wrong assumptions) remains an open question in semi-supervised learning. This means the question "which semi-supervised model should I use?" does not have an easy answer. Consequently, this book will mainly present methodology. Most chapters will introduce a distinct family of semi-supervised learning models. We start with a simple semi-supervised classification model: self-training.

2.5   SELF-TRAINING MODELS

Self-training is characterized by the fact that the learning process uses its own predictions to teach itself. For this reason, it is also called self-teaching or bootstrapping (not to be confused with the statistical procedure with the same name). Self-training can be either inductive or transductive, depending on the nature of the predictor f.

Algorithm 2.4. Self-training.

Input: labeled data $\{(\mathbf{x}_i, y_i)\}_{i=1}^{l}$, unlabeled data $\{\mathbf{x}_j\}_{j=l+1}^{l+u}$.
1. Initially, let $L = \{(\mathbf{x}_i, y_i)\}_{i=1}^{l}$ and $U = \{\mathbf{x}_j\}_{j=l+1}^{l+u}$.
2. Repeat:
3.   Train f from L using supervised learning.
4.   Apply f to the unlabeled instances in U.

---

[2]The graph-based model used here featured a Gaussian-weighted graph ($w_{ij} = \exp \frac{||x_i - x_j||^2}{2\sigma^2}$, with $\sigma = 0.1$), and predictions were made using the closed-form harmonic function solution. While this is a transductive method, we calculate the boundary as the value on the x-axis where the predicted label changes.

5.   *Remove a subset S from U; add $\{(\mathbf{x}, f(\mathbf{x})) | \mathbf{x} \in S\}$ to L.*

The main idea is to first train f on labeled data. The function f is then used to predict the labels for the unlabeled data. A subset S of the unlabeled data, together with their predicted labels, are then selected to augment the labeled data. Typically, S consists of the few unlabeled instances with the most confident f predictions. The function f is re-trained on the now larger set of labeled data, and the procedure repeats. It is also possible for S to be the whole unlabeled data set. In this case, L and U remain the whole training sample, but the assigned labels on unlabeled instances might vary from iteration to iteration.

Remark 2.5. Self-Training Assumption   The assumption of self-training is that its own predictions, at least the high confidence ones, tend to be correct. This is likely to be the case when the classes form well-separated clusters.

The major advantages of self-training are its simplicity and the fact that it is a *wrapper* method. This means that the choice of learner for f in step 3 is left completely open. For example, the learner can be a simple kNN algorithm, or a very complicated classifier. The self-training procedure "wraps" around the learner without changing its inner workings. This is important for many real world tasks like natural language processing, where the learners can be complicated black boxes not amenable to changes.

On the other hand, it is conceivable that an early mistake made by f (which is not perfect to start with, due to a small initial L) can reinforce itself by generating incorrectly labeled data. Re-training with this data will lead to an even worse f in the next iteration. Various heuristics have been proposed to alleviate this problem.

Example 2.6.   As a concrete example of self-training, we now introduce an algorithm we call *propagating 1-nearest-neighbor* and illustrate it using the little green aliens data.

Algorithm 2.7. Propagating 1-Nearest-Neighbor.

Input: labeled data $\{(\mathbf{x}_i, y_i)\}_{i=1}^{l}$, unlabeled data $\{\mathbf{x}_j\}_{j=l+1}^{l+u}$, distance function $d()$.
1. Initially, let $L = \{(\mathbf{x}_i, y_i)\}_{i=1}^{l}$ and $U = \{\mathbf{x}_j\}_{j=l+1}^{l+u}$.
2. Repeat until U is empty:
3.   Select $\mathbf{x} = \text{argmin}_{\mathbf{x} \in U} \min_{\mathbf{x}' \in L} d(\mathbf{x}, \mathbf{x}')$.
4.   Set $f(\mathbf{x})$ to the label of $\mathbf{x}$'s nearest instance in L. Break ties randomly.
5.   Remove $\mathbf{x}$ from U; add $(\mathbf{x}, f(\mathbf{x}))$ to L.

This algorithm wraps around a 1-nearest-neighbor classifier. In each iteration, it selects the unlabeled instance that is closest to any "labeled" instance (i.e., any instance currently in L, some of which were labeled by previous iterations). The algorithm approximates confidence by the distance to the currently labeled data. The selected instance is then assigned the label of its nearest neighbor and inserted into L as if it were truly labeled data. The process repeats until all instances have been added to L.

We now return to the data featuring the 100 little green aliens. Suppose you only met one male and one female alien face-to-face (i.e., labeled data), but you have unlabeled data for the weight and height of 98 others. You would like to classify all the aliens by gender, so you apply propagating 1-nearest-neighbor. Figure 2.3 illustrates the results after three particular iterations, as well as the final labeling of all instances. Note that the original labeled instances appear as large symbols, unlabeled instances as green dots, and instances labeled by the algorithm as small symbols. The figure illustrates the way the labels propagate to neighbors, expanding the sets of positive and negative instances until all instances are labeled. This approach works remarkably well and recovers the true labels exactly as they appear in Figure 1.3(a). This is because the model assumption—that the classes form well-separated clusters—is true for this data set.

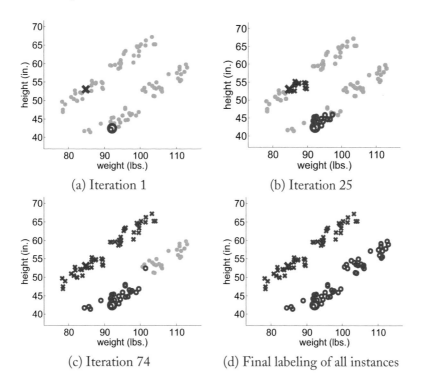

(a) Iteration 1

(b) Iteration 25

(c) Iteration 74

(d) Final labeling of all instances

Figure 2.3: Propagating 1-nearest-neighbor applied to the 100-little-green-alien data.

We now modify this data by introducing a single outlier that falls directly between the two classes. An outlier is an instance that appears unreasonably far from the rest of the data. In this case, the instance is far from the center of any of the clusters. As shown in Figure 2.4, this outlier breaks the well-separated cluster assumption and leads the algorithm astray. Clearly, self-training methods such as propagating 1-nearest-neighbor are highly sensitive to outliers that may lead to propagating incorrect information. In the case of the current example, one way to avoid this issue is to consider more than the single nearest neighbor in both selecting the next point to label as well as assigning it a label.

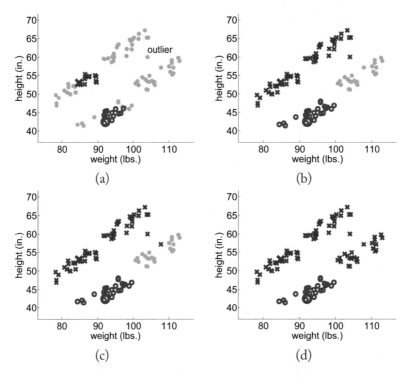

Figure 2.4: Propagating 1-nearest-neighbor illustration featuring an outlier: (a) after first few iterations, (b,c) steps highlighting the effect of the outlier, (d) final labeling of all instances, with the entire rightmost cluster mislabeled.

This concludes our basic introduction to the motivation behind semi-supervised learning, and the various issues that a practitioner must keep in mind. We also showed a simple example of semi-supervised learning to highlight the potential successes and failures. In the next chapter, we discuss in depth a more sophisticated type of semi-supervised learning algorithm that uses generative probabilistic models.

BIBLIOGRAPHICAL NOTES

Semi-supervised learning is a maturing field with extensive literature. It is impossible to cover all aspects of semi-supervised learning in this introductory book. We try to select a small sample of widely used semi-supervised learning approaches to present in the next few chapters, but have to omit many others due to space. We provide a glimpse to these other approaches in Chapter 8.

Semi-supervised learning is one way to address the scarcity of labeled data. We encourage readers to explore alternative ways to obtain labels. For example, there are ways to motivate human annotators to produce more labels via computer games [177], the sense of contribution to citizen science [165], or monetary rewards [3].

Multiple researchers have informally noted that semi-supervised learning does not *always* help. Little is written about it, except a few papers like [48, 64]. This is presumably due to "publication bias," that negative results tend not to be published. A deeper understanding of when semi-supervised learning works merits further study.

Yarowsky's word sense disambiguation algorithm [191] is a well-known early example of self training. There are theoretical analyses of self-training for specific learning algorithms [50, 80]. However, in general self-training might be difficult to analyze. Example applications of self-training can be found in [121, 144, 145].

CHAPTER 3

Mixture Models and EM

Unlabeled data tells us how the instances from *all* the classes, mixed together, are distributed. If we know how the instances from *each* class are distributed, we may decompose the mixture into individual classes. This is the idea behind *mixture models*. In this chapter, we formalize the idea of mixture models for semi-supervised learning. First we review some concepts in probabilistic modeling. Readers familiar with machine learning can skip to Section 3.2.

3.1 MIXTURE MODELS FOR SUPERVISED CLASSIFICATION

Example 3.1. Gaussian Mixture Model with Two Components   Suppose training data comes from two one-dimensional Gaussian distributions. Figure 3.1 illustrates the underlying $p(\mathbf{x}|y)$ distributions and a small training sample with only two labeled instances and several unlabeled instances.

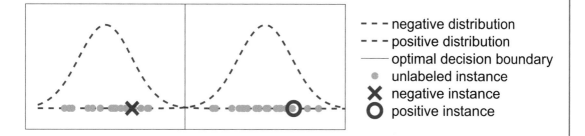

Figure 3.1: Two classes forming a mixture model with 1-dimensional Gaussian distribution components. The dashed curves are $p(\mathbf{x}|y = -1)$ and $p(\mathbf{x}|y = 1)$, respectively. The labeled and unlabeled instances are plotted on the $\mathbf{x}$-axis.

Suppose we know that the data comes from two Gaussian distributions, but we do not know their parameters (the mean, variance, and prior probabilities, which we will define soon). We can use the data (labeled and unlabeled) to estimate these parameters for both distributions. Note that, in this example, the labeled data is actually misleading: the labeled instances are both to the right of the means of the true distributions. The unlabeled data, however, helps us to identify the means of the two Gaussian distribution. Computationally, we select parameters to maximize the probability of generating such training data from the proposed model. In particular, the training samples are more likely if the means of the Gaussians are centered over the unlabeled data, rather than shifted to the right over the labeled data.

Formally, let $\mathbf{x} \in \mathcal{X}$ be an instance. We are interested in predicting its class label y. We will employ a probabilistic approach that seeks the label that maximizes the conditional probability $p(y|\mathbf{x})$. This conditional probability specifies how likely each class label is, given the instance. By definition, $p(y|\mathbf{x}) \in [0, 1]$ for all y, and $\sum_y p(y|\mathbf{x}) = 1$. If we want to minimize classification error, the best strategy is to always classify $\mathbf{x}$ into the most likely class $\hat{y}$:[1]

$$\hat{y} = \operatorname*{argmax}_y p(y|\mathbf{x}). \tag{3.1}$$

Note that if different types of misclassification (e.g., wrongly classifying a benign tumor as malignant vs. the other way around) incur different amounts of *loss*, the above strategy may not be optimal in terms of minimizing the expected loss. We defer the discussion of loss minimization to later chapters, but note that it is straightforward to handle loss in probabilistic models.

How do we compute $p(y|\mathbf{x})$? One approach is to use a *generative model*, which employs the Bayes rule:

$$p(y|\mathbf{x}) = \frac{p(\mathbf{x}|y)p(y)}{\sum_{y'} p(\mathbf{x}|y')p(y')}, \tag{3.2}$$

where the summation in the denominator is over all class labels y'. $p(\mathbf{x}|y)$ is called the *class conditional probability*, and $p(y)$ the *prior probability*. It is useful to illustrate these probability notations using the alien gender example:

- For a specific alien, $\mathbf{x}$ is the (weight, height) feature vector, and $p(y|\mathbf{x})$ is a probability distribution over two outcomes: male or female. That is, $p(y = \text{male}|\mathbf{x}) + p(y = \text{female}|\mathbf{x}) = 1$. There are infinitely many $p(y|\mathbf{x})$ distributions, one for each feature vector $\mathbf{x}$.

- There are only two class conditional distributions: $p(\mathbf{x}|y = \text{male})$ and $p(\mathbf{x}|y = \text{female})$. Each is a continuous (e.g., Gaussian) distribution over feature vectors. In other words, some weight and height combinations are more likely than others for each gender, and $p(\mathbf{x}|y)$ specifies these differences.

- The prior probabilities $p(y = \text{male})$ and $p(y = \text{female})$ specify the proportions of males and females in the alien population.

Furthermore, one can hypothetically "generate" i.i.d. instance-label pairs $(\mathbf{x}, y)$ from these probability distributions by repeating the following two steps, hence the name *generative* model:[2]

1. Sample $y \sim p(y)$. In the alien example, one can think of $p(y)$ as the probability of heads of a biased coin. Flipping the coin then selects a gender.

2. Sample $\mathbf{x} \sim p(\mathbf{x}|y)$. In the alien example, this samples a two-dimensional feature vector to describe an alien of the gender chosen in step 1.

---

[1]Note that a "hat" on a variable (e.g., $\hat{y}, \hat{\theta}$) indicates we are referring to an estimated or predicted value.

[2]An alternative to generative models are *discriminative* models, which focus on distinguishing the classes without worrying about the process underlying the data generation.

We call $p(\mathbf{x}, y) = p(y)p(\mathbf{x}|y)$ the *joint distribution* of instances and labels.

As an example of generative models, the multivariate Gaussian distribution is a common choice for continuous feature vectors $\mathbf{x}$. The class conditional distributions have the probability density function

$$p(\mathbf{x}|y) = \mathcal{N}(\mathbf{x}; \mu_y, \Sigma_y) = \frac{1}{(2\pi)^{D/2}|\Sigma_y|^{1/2}} \exp\left(-\frac{1}{2}(\mathbf{x} - \mu_y)^\top \Sigma_y^{-1}(\mathbf{x} - \mu_y)\right), \qquad (3.3)$$

where μ_y and Σ_y are the mean vector and covariance matrix, respectively, An example task is image classification, where $\mathbf{x}$ may be the vector of pixel intensities of an image. Images in each class are modeled by a Gaussian distribution. The overall generative model is called a Gaussian Mixture Model (GMM).

As another example of generative models, the multinomial distribution

$$p(\mathbf{x} = (x_{.1}, \ldots, x_{.d})|\mu_y) = \frac{(\sum_{i=1}^{D} x_{.i})!}{x_{.1}! \cdots x_{.D}!} \prod_{d=1}^{D} \mu_{yd}^{x_{.d}}, \qquad (3.4)$$

where μ_y is a probability vector, is a common choice for modeling count vectors x. For instance, in text categorization $\mathbf{x}$ is the vector of word counts in a document (the so-called bag-of-words representation). Documents in each category are modeled by a multinomial distribution. The overall generative model is called a Multinomial Mixture Model.

As yet another example of generative models, Hidden Markov Models (HMM) are commonly used to model *sequences* of instances. Each instance in the sequence is generated from a hidden state, where the state conditional distribution can be a Gaussian or a multinomial, for example. In addition, HMMs specify the transition probability between states to form the sequence. Learning HMMs involves estimating the conditional distributions' parameters and transition probabilities. Doing so makes it possible to infer the hidden states responsible for generating the instances in the sequences.

Now we know how to do classification once we have $p(\mathbf{x}|y)$ and $p(y)$, but the problem remains to learn these distributions from training data. The class conditional $p(\mathbf{x}|y)$ is often determined by some model parameters, for example, the mean μ and covariance matrix Σ of a Gaussian distribution. For $p(y)$, if there are C classes we need to estimate $C - 1$ parameters: $p(y = 1), \ldots, p(y = C - 1)$. The probability $p(y = C)$ is constrained to be $1 - \sum_{c=1}^{C-1} p(y = c)$ since $p(y)$ is normalized. We will use θ to denote the set of all parameters in $p(\mathbf{x}|y)$ and $p(y)$. If we want to be explicit, we use the notation $p(\mathbf{x}|y, \theta)$ and $p(y|\theta)$. Training amounts to finding a good θ. But how do we define goodness?

One common criterion is the *maximum likelihood estimate* (MLE). Given training data $\mathcal{D}$, the MLE is

$$\hat{\theta} = \operatorname*{argmax}_\theta p(\mathcal{D}|\theta) = \operatorname*{argmax}_\theta \log p(\mathcal{D}|\theta). \qquad (3.5)$$

That is, the MLE is the parameter under which the data likelihood $p(\mathcal{D}|\theta)$ is the largest. We often work with log likelihood $\log p(\mathcal{D}|\theta)$ instead of the straight likelihood $p(\mathcal{D}|\theta)$. They yield the same maxima since $\log()$ is monotonic, and log likelihood will be easier to handle.

In supervised learning when $\mathcal{D} = \{(\mathbf{x}_i, y_i)\}_{i=1}^{l}$, the MLE is usually easy to find. We can rewrite the log likelihood as

$$\log p(\mathcal{D}|\theta) = \log \prod_{i=1}^{l} p(\mathbf{x}_i, y_i|\theta) = \sum_{i=1}^{l} \log p(y_i|\theta) p(\mathbf{x}_i|y_i, \theta), \tag{3.6}$$

where we used the fact that the probability of a set of i.i.d. events is the product of individual probabilities. Finding an MLE is an optimization problem to maximize the log likelihood. In supervised learning, the optimization problem is often straightforward and yields intuitive MLE solutions, as the next example shows.

Example 3.2. MLE for Gaussian Mixture Model, All Labeled Data   We now present the derivation for the maximum likelihood estimate for a 2-class Gaussian mixture model when $\mathcal{D} = \{(\mathbf{x}_i, y_i)\}_{i=1}^{l}$. We begin by setting up the constrained optimization problem

$$\hat{\theta} = \underset{\theta}{\operatorname{argmax}} \log p(\mathcal{D}|\theta) \quad \text{s.t.} \sum_{j=1}^{2} p(y_j|\theta) = 1, \tag{3.7}$$

where we enforce the constraint that the class priors must sum to 1. We next introduce a Lagrange multiplier β to form the Lagrangian (see [99] for a tutorial on Lagrange multipliers)

$$
\begin{aligned}
\Lambda(\theta, \beta) &= \log p(\mathcal{D}|\theta) - \beta(\sum_{j=1}^{2} p(y_j|\theta) - 1) \\
&= \log \prod_{i=1}^{l} p(\mathbf{x}_i, y_i|\theta) - \beta(\sum_{j=1}^{2} p(y_j|\theta) - 1) \\
&= \sum_{i=1}^{l} \log p(y_i|\theta) p(\mathbf{x}_i|y_i, \theta) - \beta(\sum_{j=1}^{2} p(y_j|\theta) - 1) \\
&= \sum_{i=1}^{l} \log \pi_i + \sum_{i=1}^{l} \log \mathcal{N}(\mathbf{x}_i; \mu_{y_i}, \Sigma_{y_i}) - \beta(\sum_{j=1}^{2} \pi_j - 1),
\end{aligned}
$$

where π_j, μ_j, Σ_j for $j \in \{1, 2\}$ are the class priors and Gaussian means and covariance matrices. We compute the partial derivatives with respect to all the parameters. We then set each partial derivative to zero to obtain the intuitive closed-form MLE solution:

$$\frac{\partial \Lambda}{\partial \beta} = \sum_{j=1}^{2} \pi_j - 1 = 0 \Rightarrow \sum_{j=1}^{2} \pi_j = 1. \tag{3.8}$$

Clearly, the β Lagrange multiplier's role is to enforce the normalization constraint on the class priors.

$$\frac{\partial \Lambda}{\partial \pi_j} = \sum_{i:y_i=j} \frac{1}{\pi_j} - \beta = \frac{l_j}{\pi_j} - \beta = 0 \Rightarrow \pi_j = \frac{l_j}{\beta} = \frac{l_j}{l}, \tag{3.9}$$

where l_j is the number of instances in class j. Note that we find $\beta = l$ by substituting (3.9) into (3.8) and rearranging. In the end, we see that the MLE for the class prior is simply the fraction of instances in each class. We next solve for the class means. For this we need to use the derivatives with respect to the vector μ_j. In general, let v be a vector and A a square matrix of the appropriate size, we have $\frac{\partial}{\partial v} v^\top A v = 2 A v$. This leads to

$$
\begin{aligned}
\frac{\partial \Lambda}{\partial \mu_j} &= \frac{\partial}{\partial \mu_j} \sum_{i:y_i=j} -\frac{1}{2}(\mathbf{x}_i - \mu_j)^\top \Sigma_j^{-1}(\mathbf{x}_i - \mu_j) \\
&= \sum_{i:y_i=j} \Sigma_j^{-1}(\mathbf{x}_i - \mu_j) = 0 \Rightarrow \mu_j = \frac{1}{l_j} \sum_{i:y_i=j} \mathbf{x}_i .
\end{aligned}
\tag{3.10}
$$

We see that the MLE for each class mean is simply the class's sample mean. Finally, the MLE solution for the covariance matrices is

$$
\Sigma_j = \frac{1}{l_j} \sum_{i:y_i=j} (\mathbf{x}_i - \mu_j)(\mathbf{x}_i - \mu_j)^\top ,
\tag{3.11}
$$

which is the sample covariance for the instances of that class.

3.2 MIXTURE MODELS FOR SEMI-SUPERVISED CLASSIFICATION

In semi-supervised learning, $\mathcal{D}$ consists of both labeled and unlabeled data. The likelihood depends on both the labeled and unlabeled data—this is how unlabeled data might help semi-supervised learning in mixture models. It is no longer possible to solve the MLE analytically. However, as we will see in the next section, one can find a local maximum of the parameter estimate using an iterative procedure known as the EM algorithm.

Since the training data consists of both labeled and unlabeled data, i.e., $\mathcal{D} = \{(\mathbf{x}_1, y_1), \ldots, (\mathbf{x}_l, y_l), \mathbf{x}_{l+1}, \ldots, \mathbf{x}_{l+u}\}$, the log likelihood function is now defined as

$$
\log p(\mathcal{D}|\theta) = \log \left(\prod_{i=1}^{l} p(\mathbf{x}_i, y_i|\theta) \prod_{i=l+1}^{l+u} p(\mathbf{x}_i|\theta) \right)
\tag{3.12}
$$

$$
= \sum_{i=1}^{l} \log p(y_i|\theta) p(\mathbf{x}_i|y_i, \theta) + \sum_{i=l+1}^{l+u} \log p(\mathbf{x}_i|\theta) .
\tag{3.13}
$$

The essential difference between this semi-supervised log likelihood (3.13) and the previous supervised log likelihood (3.6) is the second term for unlabeled instances. We call $p(\mathbf{x}|\theta)$ the *marginal probability*, which is defined as

$$
p(\mathbf{x}|\theta) = \sum_{y=1}^{C} p(\mathbf{x}, y|\theta) = \sum_{y=1}^{C} p(y|\theta) p(\mathbf{x}|y, \theta) .
\tag{3.14}
$$

It is the probability of generating $\mathbf{x}$ from any of the classes. The marginal probabilities therefore account for the fact that we know which unlabeled instances are present, but not which classes they belong to. Semi-supervised learning in mixture models amounts to finding the MLE of (3.13). The only difference between supervised and semi-supervised learning (for mixture models) is the objective function being maximized. Intuitively, a semi-supervised MLE $\hat{\theta}$ will need to fit both the labeled and unlabeled instances. Therefore, we can expect it to be different from the supervised MLE.

The unobserved labels $y_{l+1}, \ldots, y_{l+u}$ are called hidden variables. Unfortunately, hidden variables can make the log likelihood (3.13) non-concave and hard to optimize.[3] Fortunately, there are several well studied optimization methods which attempt to find a locally optimal θ. We will present a particularly important method, the *Expectation Maximization* (EM) algorithm, in the next section. For Gaussian Mixture Models (GMMs), Multinomial Mixture Models, HMMs, etc., the EM algorithm has been the *de facto* standard optimization technique to find an MLE when unlabeled data is present. The EM algorithm for HMMs even has its own name: the Baum-Welch algorithm.

3.3   OPTIMIZATION WITH THE EM ALGORITHM*

Recall the observed data is $\mathcal{D} = \{(\mathbf{x}_1, y_1), \ldots, (\mathbf{x}_l, y_l), \mathbf{x}_{l+1}, \ldots, \mathbf{x}_{l+u}\}$, and let the hidden data be $\mathcal{H} = \{y_{l+1}, \ldots, y_{l+u}\}$. Let the model parameter be θ. The EM algorithm is an iterative procedure to find a θ that locally maximizes $p(\mathcal{D}|\theta)$.

Algorithm 3.3. The Expectation Maximization (EM) Algorithm in General.

Input: observed data $\mathcal{D}$, hidden data $\mathcal{H}$, initial parameter $\theta^{(0)}$
1. Initialize $t = 0$.
2. Repeat the following steps until $p(\mathcal{D}|\theta^{(t)})$ converges:
3.   E-step: compute $q^{(t)}(\mathcal{H}) \equiv p(\mathcal{H}|\mathcal{D}, \theta^{(t)})$
4.   M-step: find $\theta^{(t+1)}$ that maximizes $\sum_{\mathcal{H}} q^{(t)}(\mathcal{H}) \log p(\mathcal{D}, \mathcal{H}|\theta^{(t+1)})$
5. $t = t + 1$
Output: $\theta^{(t)}$

We comment on a few important aspects of the EM algorithm:

- $q^{(t)}(\mathcal{H})$ is the hidden label distribution. It can be thought of as assigning "soft labels" to the unlabeled data according to the current model $\theta^{(t)}$.

- It can be proven that EM improves the log likelihood $\log p(\mathcal{D}|\theta)$ at each iteration. However, EM only converges to a *local optimum*. That is, the θ it finds is only guaranteed to be the best within a neighborhood in the parameter space; θ may not be a *global optimum* (the desired

---

[3]The reader is invited to form the Lagrangian and solve the zero partial derivative equations. The parameters will end up on both sides of the equation, and there will be no analytical solution for the MLE $\hat{\theta}$.

MLE). The complete proof is beyond the scope of this book and can be found in standard textbooks on EM. See the references in the bibliographical notes.

- The local optimum to which EM converges depends on the initial parameter $\theta^{(0)}$. A common choice of $\theta^{(0)}$ is the MLE on the small labeled training data.

The above general EM algorithm needs to be specialized for specific generative models.

Example 3.4. EM for a 2-class GMM with Hidden Variables We illustrate the EM algorithm on a simple GMM generative model. In this special case, the observed data is simply a sample of labeled and unlabeled instances. The hidden variables are the labels of the unlabeled instances. We learn the parameters of two Gaussians to fit the data using EM as follows:

Algorithm 3.5. EM for GMM.

Input: observed data $\mathcal{D} = \{(\mathbf{x}_1, y_1), \ldots, (\mathbf{x}_l, y_l), \mathbf{x}_{l+1}, \ldots, \mathbf{x}_{l+u}\}$

1. *Initialize* $t = 0$ *and* $\theta^{(0)} = \{\pi_j^{(0)}, \mu_j^{(0)}, \Sigma_j^{(0)}\}_{j \in \{1,2\}}$ *to the MLE estimated from labeled data.*
2. *Repeat until the log likelihood* $\log p(\mathcal{D}|\theta)$ *converges:*
3.   *E-step: For all unlabeled instances* $i \in \{l+1, \ldots, l+u\}$, $j \in \{1, 2\}$, *compute*

$$\gamma_{ij} \equiv p(y_j|\mathbf{x}_i, \theta^{(t)}) = \frac{\pi_j^{(t)} \mathcal{N}(\mathbf{x}_i; \mu_j^{(t)}, \Sigma_j^{(t)})}{\sum_{k=1}^{2} \pi_k^{(t)} \mathcal{N}(\mathbf{x}_i; \mu_k^{(t)}, \Sigma_k^{(t)})}. \tag{3.15}$$

   For labeled instances, define $\gamma_{ij} = 1$ *if* $y_i = j$, *and 0 otherwise.*
4.   *M-step: Find* $\theta^{(t+1)}$ *using the current* γ_{ij}. *For* $j \in \{1, 2\}$,

$$l_j = \sum_{i=1}^{l+u} \gamma_{ij} \tag{3.16}$$

$$\mu_j^{(t+1)} = \frac{1}{l_j} \sum_{i=1}^{l+u} \gamma_{ij} \mathbf{x}_i \tag{3.17}$$

$$\Sigma_j^{(t+1)} = \frac{1}{l_j} \sum_{i=1}^{l+u} \gamma_{ij} (\mathbf{x}_i - \mu_j^{(t+1)})(\mathbf{x}_i - \mu_j^{(t+1)})^{\top} \tag{3.18}$$

$$\pi_j^{(t+1)} = \frac{l_j}{l+u} \tag{3.19}$$

5. $t = t + 1$
Output: $\{\pi_j, \mu_j, \Sigma_j\}_{j=\{1,2\}}$

Note that the algorithm begins by finding the MLE of the labeled data alone. The E-step then computes the γ values, which can be thought of as soft assignments of class labels to instances. These are sometimes referred to as "responsibilities" (i.e., class 1 is responsible for x_i with probability γ_{i1}).

The M-step updates the model parameters using the current γ values as weights on the unlabeled instances. If we think of the E-step as creating fractional labeled instances split between the classes, then the M-step simply computes new MLE parameters using these fractional instances and the labeled data. The algorithm stops when the log likelihood (3.13) converges (i.e., stops changing from one iteration to the next). The data log likelihood in the case of a mixture of two Gaussians is

$$\log p(\mathcal{D}|\theta) \;=\; \sum_{i=1}^{l} \log \pi_{y_i} \mathcal{N}(\mathbf{x}_i; \mu_{y_i}, \Sigma_{y_i}) + \sum_{i=l+1}^{l+u} \log \sum_{j=1}^{2} \pi_j \mathcal{N}(\mathbf{x}_i; \mu_j, \Sigma_j), \quad (3.20)$$

where we have marginalized over the two classes for the unlabeled data.

It is instructive to note the similarity between EM and self-training. EM can be viewed as a special form of self-training, where the current classifier θ would label the unlabeled instances with all possible labels, but each with *fractional weights* $p(\mathcal{H}|\mathcal{D}, \theta)$. Then *all* these augmented unlabeled data, instead of the top few most confident ones, are used to update the classifier.

3.4   THE ASSUMPTIONS OF MIXTURE MODELS

Mixture models provide a framework for semi-supervised learning in which the role of unlabeled data is clear. In practice, this form of semi-supervised learning can be highly effective if the generative model is (nearly) correct. It is worth noting the assumption made here:

Remark 3.6. Mixture Model Assumption   The data actually comes from the mixture model, where the number of components, prior $p(y)$, and conditional $p(\mathbf{x}|y)$ are all correct.

Unfortunately, it can be difficult to assess the model correctness since we do not have much labeled data. Many times one would choose a generative model based on domain knowledge and/or mathematical convenience. However, if the model is wrong, semi-supervised learning could actually hurt performance. In this case, one might be better off to use only the labeled data and perform supervised learning instead. The following example shows the effect of an incorrect model.

Example 3.7. An Incorrect Generative Model   Suppose a dataset contains four clusters of data, two of each class. This dataset is shown in Figure 3.2. The correct decision boundary is a horizontal line along the x-axis. Clearly, the data is not generated from two Gaussians. If we insist that each class is modeled by a single Gaussian, the results may be poor. Figure 3.3 illustrates this point by comparing two possible GMMs fitting this data. In panel (a), the learned model fits the unlabeled quite well (having high log likelihood), but predictions using this model will result in approximately 50% error. In contrast, the model shown in panel (b) will lead to much better accuracy. However, (b) would not be favored by the EM algorithm since it has a lower log likelihood.

As mentioned above, we may be better off using only labeled data and supervised learning in this case. If we have labeled data in the bottom left cluster and top right cluster, the supervised

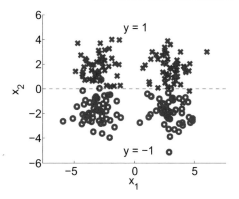

Figure 3.2: Two classes in four clusters (each a 2-dimensional Gaussian distribution).

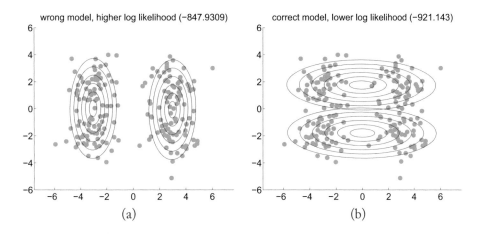

Figure 3.3: (a) Good fit under the wrong model assumption. The decision boundary is vertical, thus producing mass misclassification. (b) Worse fit under the wrong model assumption. However, the decision boundary is correct.

decision boundary would be approximately the line $y = -x$, which would result in only about 25% error.

There are a number of ways to alleviate the danger of using the wrong model. One obvious way is to refine the model to fit the task better, which requires domain knowledge. In the above example, one might model each class itself as a GMM with two components, instead of a single Gaussian.

Another way is to de-emphasize the unlabeled data, in case the model correctness is uncertain. Specifically, we scale the contribution from unlabeled data in the semi-supervised log likelihood (3.13) by a small positive weight $\lambda < 1$:

$$\sum_{i=1}^{l} \log p(y_i|\theta) p(\mathbf{x}_i|y_i, \theta) + \lambda \sum_{i=l+1}^{l+u} \log p(\mathbf{x}_i|\theta). \tag{3.21}$$

As $\lambda \to 0$, the influence of unlabeled data vanishes and one recovers the supervised learning objective.

3.5 OTHER ISSUES IN GENERATIVE MODELS

When defining a generative model, *identifiability* is a desirable property. A model is identifiable if $p(x|\theta_1) = p(x|\theta_2) \iff \theta_1 = \theta_2$, up to a permutation of mixture component indices. That is, two models are considered equivalent if they differ only by which component is called component one, which is called component two, and so on. That is to say, there is a unique (up to permutation) model θ that explains the observed unlabeled data. Therefore, as the size of unlabeled data grows, one can hope to accurately recover the mixing components. For instance, GMMs are identifiable, while some other models are not. The following example shows an unidentifiable model and why it is not suitable for semi-supervised learning.

Example 3.8. An Unidentifiable Generative Model   Assume the component model $p(x|y)$ is uniform for $y \in \{+1, -1\}$. Let us try to use semi-supervised learning to learn the mixture of uniform distributions. We are given a large amount of unlabeled data, such that we know $p(x)$ is uniform in $[0, 1]$. We also have 2 labeled data points $(0.1, -1)$, $(0.9, +1)$. Can we determine the label for $x = 0.5$?

The answer turns out to be no. With our assumptions, we cannot distinguish the following two models (and infinitely many others):

$$p(y = -1) = 0.2, \ p(x|y = -1) = \text{unif}(0, 0.2), \ p(x|y = 1) = \text{unif}(0.2, 1) \tag{3.22}$$
$$p(y = -1) = 0.6, \ p(x|y = -1) = \text{unif}(0, 0.6), \ p(x|y = 1) = \text{unif}(0.6, 1) \tag{3.23}$$

Both models are consistent with the unlabeled data and the labeled data, but the first model predicts label $y = 1$ at $x = 0.5$, while the second model predicts $y = -1$. This is illustrated by Figure 3.4.

Another issue with generative models is *local optima*. Even if the model is correct and identifiable, the log likelihood (3.13) as a function of model parameters θ is, in general, non-concave. That is, there might be multiple "bumps" on the surface. The highest bump corresponds to the *global optimum*, i.e., the desired MLE. The other bumps are local optima. The EM algorithm is prone to being trapped in a local optimum. Such local optima might lead to inferior performance. A standard practice against local optima is *random restart*, in which the EM algorithm is run multiple times. Each time EM starts from a different random initial parameter $\theta^{(0)}$. Finally, the log likelihood that EM converges to in each run is compared. The θ that correspond to the best log likelihood is selected. It is worth noting that random restart does not solve the local optima problem—it only alleviates

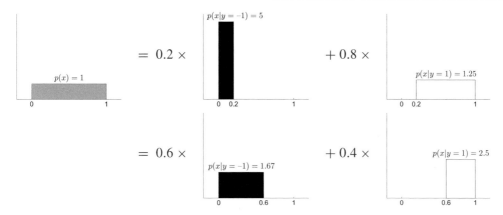

Figure 3.4: An example of unidentifiable models. Even if we know $p(x)$ is a mixture of two uniform distributions, we cannot uniquely identify the two components. For instance, the two mixtures produce the same $p(x)$, but they classify $x = 0.5$ differently. Note the height of each distribution represents a probability density (which can be greater than 1), not probability mass. The area under each distribution is 1.

it. Selecting a better $\theta^{(0)}$ that is more likely to lead to the global optimum (or simply a better local optimum) is another heuristic method, though this may require domain expertise.

Finally, we note that the goal of optimization for semi-supervised learning with mixture models is to maximize the log likelihood (3.13). The EM algorithm is only one of several optimization methods to find a (local) optimum. Direct optimization methods are possible, too, for example quasi-Newton methods like L-BFGS [115].

3.6 CLUSTER-THEN-LABEL METHODS

We have used the EM algorithm to identify the mixing components from unlabeled data. Recall that unsupervised clustering algorithms can also identify clusters from unlabeled data. This suggests a natural *cluster-then-label* algorithm for semi-supervised classification.

Algorithm 3.9. Cluster-then-Label.

Input: labeled data $(\mathbf{x}_1, y_1), \ldots, (\mathbf{x}_l, y_l)$, *unlabeled data* $\mathbf{x}_{l+1}, \ldots, \mathbf{x}_{l+u}$,
    a clustering algorithm $\mathcal{A}$, *and a supervised learning algorithm* $\mathcal{L}$
1. *Cluster* $\mathbf{x}_1, \ldots, \mathbf{x}_{l+u}$ *using* $\mathcal{A}$.
2. *For each resulting cluster, let S be the labeled instances in this cluster:*
3.   *If S is non-empty, learn a supervised predictor from S:* $f_S = \mathcal{L}(S)$.
       Apply f_S *to all unlabeled instances in this cluster.*
4.   *If S is empty, use the predictor* f *trained from all labeled data.*
Output: labels on unlabeled data $y_{l+1}, \ldots, y_{l+u}$.

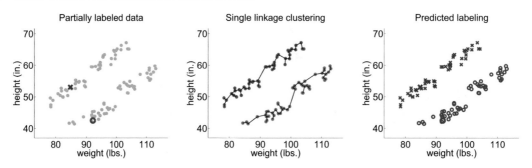

Figure 3.5: Cluster-then-label results using single linkage hierarchical agglomerative clustering ($\mathcal{A}$) and majority vote ($\mathcal{L}$).

In step 1, the clustering algorithm $\mathcal{A}$ is unsupervised. In step 2, we learn one supervised predictor using the labeled instances that fall into each cluster, and use the predictor to label the unlabeled instances in that cluster. One can use any clustering algorithm $\mathcal{A}$ and supervised learner $\mathcal{L}$. It is worth noting that cluster-then-label does not necessarily involve a probabilistic mixture model.

The following example shows cluster-then-label with Hierarchical Agglomerative Clustering for $\mathcal{A}$, and simple majority vote within each cluster for $\mathcal{L}$.

Example 3.10. Cluster-then-Label with Hierarchical Agglomerative Clustering   In this example, we apply the cluster-then-label approach to the aliens data from Chapter 1. For step 1, we use hierarchical agglomerative clustering (Algorithm 1.1) with Euclidean distance as the underlying distance function, and the single linkage method to determine distances between clusters. Because the labeled data contains only two classes, we heuristically stop the algorithm once only two clusters remain. Steps 2–4 of cluster-then-label find the majority label within each cluster, and assign this label to all unlabeled instances in the cluster.

Figure 3.5 shows the original partially labeled data, the two clusters, and the final labels predicted for all the data. In this case, because the clusters coincide with the true labeling of the data, we correctly classify all unlabeled instances. The lines between instances in the center panel indicate the linkages responsible for merging clusters until only two remain.

It turns out that using single linkage is critically important here, where the natural clusters are long and skinny. A different choice—e.g., complete linkage clustering—tends to form rounder clusters. When applied to this data, complete linkage produces the results shown in Figure 3.6. The clusters found by complete linkage do not match up with the true labeling of the data. In fact, both labeled instances end up in the same cluster. Because there is no majority label in either cluster, the majority vote algorithm ends up breaking ties randomly, thus assigning random labels to all unlabeled instances.

The point of this example is *not* to show that complete linkage is bad and single linkage is good. In fact, it could have been the other way around for other datasets! Instead, the example is

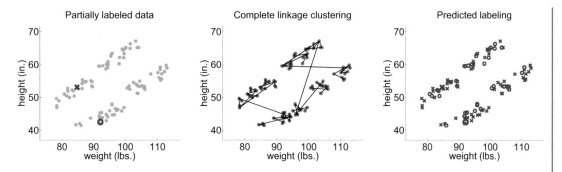

Figure 3.6: Cluster-then-label results using complete linkage hierarchical agglomerative clustering. This clustering result does not match the true labeling of the data.

meant to highlight the sensitivity of semi-supervised learning to its underlying assumptions—in this case, that the clusters coincide with decision boundaries. If this assumption is incorrect, the results can be poor.

This chapter introduced mixture models and the expectation maximization (EM) algorithm for semi-supervised learning. We also reviewed some of the common issues faced when using generative models. Finally, we presented a non-probabilistic, cluster-then-label approach using the same intuition behind mixture models: the unlabeled data helps identify clusters in the input space that correspond to each class. In the next chapter, we turn to a different semi-supervised learning approach known as co-training, which uses a very different intuition involving multiple feature representations of instances.

BIBLIOGRAPHICAL NOTES

The theoretical value of labeled and unlabeled data in the context of parametric mixture models has been analyzed as early as in [30, 142]. Under certain conditions [62, 161], theoretic analysis also justifies the Cluster-then-Label procedure [59, 52, 74]. It has also been noted that if the mixture model assumption is wrong, unlabeled data can in fact hurt performance [48].

In a seminal empirical paper [135], Nigam *et al.* applied mixture of multinomial distributions for semi-supervised learning to the task of text document categorization. Since then, similar algorithms have been successfully applied to other tasks [13, 66, 67]. Some variations, which use more than one mixture components per class, or down-weight unlabeled data relative to labeled data, can be found in [28, 43, 128, 135, 152].

The EM algorithm was originally described in [60]. More recent interpretations can be found in, e.g., [19]. Some discussions on identifiability in the context of semi-supervised learning can be found in [43, 125, 142]. Local optima issues can be addressed by smart choice of starting point using active learning [133].

CHAPTER 4

Co-Training

4.1 TWO VIEWS OF AN INSTANCE

Consider the supervised learning task of *named entity classification* in natural language processing. A named entity is a proper name such as "Washington State" or "Mr. Washington." Each named entity has a class label depending on what it is referring to. For simplicity, we assume there are only two classes: `Person` or `Location`. The goal of named entity classification is to assign the correct label to each entity, for example, `Location` to "Washington State" and `Person` to "Mr. Washington." Named entity classification is obviously a classification problem, to predict the class y from the features $\mathbf{x}$. Our focus is not on the details of training supervised classifiers that work on strings. (Basically, it involves some form of partial string matching. The details can be found in the bibliographical notes.) Instead, we focus on named entity classification as one example task that involves instances with a special structure that lends itself well to semi-supervised learning.

An instance of a named entity can be represented by two distinct sets of features. The first is the set of words that make up the named entity itself. The second is the set of words in the *context* in which the named entity occurs. In the following examples the named entity is in parentheses, and the context is underlined:

> instance 1:   ... headquartered in (Washington State) ...
> instance 2:   ... (Mr. Washington), the vice president of ...

Formally, each named entity instance is represented by two *views* (sets of features): the words in itself $\mathbf{x}^{(1)}$, and the words in its context $\mathbf{x}^{(2)}$. We write $\mathbf{x} = [\mathbf{x}^{(1)}, \mathbf{x}^{(2)}]$.

As another example of views, consider Web page classification into `Student` or `Faculty` Web pages. In this task, the first view $\mathbf{x}^{(1)}$ can be the words on the Web page in question. The second view $\mathbf{x}^{(2)}$ can be the words in all the hyperlinks that point to the Web page.

Going back to the named entity classification task, let us assume we only have these two labeled instances in our training data:

instance	$\mathbf{x}^{(1)}$	$\mathbf{x}^{(2)}$	y
1.	Washington State	headquartered in	Location
2.	Mr. Washington	vice president	Person

This labeled training sample seems woefully inadequate: we know that there are many other ways to express a location or person. For example,

> ... (Robert Jordan), a partner at ...
> ... flew to (China) ...

Because these latter instances are not covered by the two labeled instances in our training sample, a supervised learner will not be able to classify them correctly. It seems that a very large *labeled* training sample is necessary to cover all the variations in location or person expressions. Or is it?

4.2   CO-TRAINING

It turns out that one does not need a large labeled training sample for this task. It is sufficient to have a large *unlabeled* training sample, which is much easier to obtain. Let us say we have the following unlabeled instances:

instance 3:   . . . headquartered in (Kazakhstan) . . .
instance 4:   . . . flew to (Kazakhstan) . . .
instance 5:   . . . (Mr. Smith), a partner at Steptoe & Johnson . . .

It is illustrative to inspect the features of the labeled and unlabeled instances together:

instance	$\mathbf{x}^{(1)}$	$\mathbf{x}^{(2)}$	y
1.	Washington State	headquartered in	Location
2.	Mr. Washington	vice president	Person
3.	Kazakhstan	headquartered in	?
4.	Kazakhstan	flew to	?
5.	Mr. Smith	partner at	?

One may reason about the data in the following steps:

1. From labeled instance 1, we learn that "headquartered in" is a context that seems to indicate $y =$ Location.

2. If this is true, we infer that "Kazakhstan" must be a Location since it appears with the same context "headquartered in" in instance 3.

3. Since instance 4 is also about "Kazakhstan," it follows that its context "flew to" should indicate Location.

4. At this point, we are able to classify "China" in "flew to (China)" as a Location, even though neither "flew to" nor "China" appeared in the labeled data!

5. Similarly, by matching "Mr. *" in instances 2 and 5, we learn that "partner at" is a context for $y =$ Person. This allows us to classify "(Robert Jordan), a partner at" as Person, too.

This process bears a strong resemblance to the self-training algorithm in Section 2.5, where a classifier uses its most confident predictions on unlabeled instances to teach itself. There is a critical difference, though: we implicitly used *two* classifiers in turn. They operate on different views of an instance: one is based on the named entity string itself ($\mathbf{x}^{(1)}$), and the other is based on the context string ($\mathbf{x}^{(2)}$). The two classifiers *teach each other*. One can formalize this process into a *Co-Training* algorithm.

Algorithm 4.1. Co-Training.

Input: labeled data $\{(\mathbf{x}_i, y_i)\}_{i=1}^{l}$, unlabeled data $\{\mathbf{x}_j\}_{j=l+1}^{l+u}$, a learning speed k.
    Each instance has two views $\mathbf{x}_i = [\mathbf{x}_i^{(1)}, \mathbf{x}_i^{(2)}]$.
1. Initially let the training sample be $L_1 = L_2 = \{(\mathbf{x}_1, y_1), \ldots, (\mathbf{x}_l, y_l)\}$.
2. Repeat until unlabeled data is used up:
3.   Train a view-1 classifier $f^{(1)}$ from L_1, and a view-2 classifier $f^{(2)}$ from L_2.
4.   Classify the remaining unlabeled data with $f^{(1)}$ and $f^{(2)}$ separately.
5.   Add $f^{(1)}$'s top k most-confident predictions $(\mathbf{x}, f^{(1)}(\mathbf{x}))$ to L_2.
         Add $f^{(2)}$'s top k most-confident predictions $(\mathbf{x}, f^{(2)}(\mathbf{x}))$ to L_1.
         Remove these from the unlabeled data.

Note $f^{(1)}$ is a view-1 classifier: although we give it the complete feature $\mathbf{x}$, it only pays attention to the first view $\mathbf{x}^{(1)}$ and ignores the second view $\mathbf{x}^{(2)}$. $f^{(2)}$ is the other way around. They each provide their most confident unlabeled-data predictions as the training data for the other view. In this process, the unlabeled data is eventually exhausted.

Co-training is a wrapper method. That is to say, it does not matter what the learning algorithms are for the two classifiers $f^{(1)}$ and $f^{(2)}$. The only requirement is that the classifiers can assign a confidence score to their predictions. The confidence score is used to select which unlabeled instances to turn into additional training data for the other view. Being a wrapper method, Co-Training is widely applicable to many tasks.

4.3   THE ASSUMPTIONS OF CO-TRAINING

Co-Training makes several assumptions. The most obvious one is the existence of two separate views $\mathbf{x} = [\mathbf{x}^{(1)}, \mathbf{x}^{(2)}]$. For a general task, the features may not naturally split into two views. To apply Co-Training in this case, one can randomly split the features into two artificial views. Assuming there are two views, the success of Co-Training depends on the following two assumptions:

Remark 4.2. Co-Training Assumptions

1. Each view alone is sufficient to make good classifications, given enough labeled data.

2. The two views are *conditionally independent* given the class label.

The first assumption is easy to understand. It not only requires that there are two views, but two good ones. The second assumption is subtle but strong. It states that

$$
\begin{aligned}
P(\mathbf{x}^{(1)}|y, \mathbf{x}^{(2)}) &= P(\mathbf{x}^{(1)}|y) \\
P(\mathbf{x}^{(2)}|y, \mathbf{x}^{(1)}) &= P(\mathbf{x}^{(2)}|y).
\end{aligned} \tag{4.1}
$$

In other words, if we know the true label y, then knowing one view (e.g., $\mathbf{x}^{(2)}$) does not affect what we will observe for the other view (it will simply be $P(\mathbf{x}^{(1)}|y)$). To illustrate the second assumption, consider our named entity classification task again. Let us collect all instances with true label $y =$ Location. View 1 of these instances will be Location named entity strings, i.e., $\mathbf{x}^{(1)} \in$ {Washington State, Kazakhstan, China, . . .}. The frequency of observing these named entities, given $y =$ Location, is described by $P(\mathbf{x}^{(1)}|y)$. These named entities are associated with various contexts. Now let us select *any* particular context, say $\mathbf{x}^{(2)} =$ "headquartered in," and consider the instances with this context and $y =$ Location. If conditional independence holds, in these instances we will again find all those named entities {Washington State, Kazakhstan, China, . . .} with the same frequencies as indicated by $P(\mathbf{x}^{(1)}|y)$. In other words, the context "headquartered in" does not favor any particular location.

Why is the conditional independence assumption important for Co-Training? If the view-2 classifier $f^{(2)}$ decides that the context "headquartered in" indicates Location with high confidence, Co-Training will add unlabeled instances with that context as view-1 training examples. These new training examples for $f^{(1)}$ will include all representative Location named entities $\mathbf{x}^{(1)}$, thanks to the conditional independence assumption. If the assumption didn't hold, the new examples could all be highly similar and thus be less informative for the view-1 classifier. It can be shown that if the two assumptions hold, Co-Training can learn successfully from labeled and unlabeled data. However, it is actually difficult to find tasks in practice that completely satisfy the conditional independence assumption. After all, the context "Prime Minister of" practically rules out most locations except countries. When the conditional independence assumption is violated, Co-Training may not perform well.

There are several variants of Co-Training. The original Co-Training algorithm picks the top k most confident unlabeled instances in each view, and augments them with predicted labels. In contrast, the so-called Co-EM algorithm is less categorical. Co-EM maintains a probabilistic model $P(y|\mathbf{x}^{(v)}; \theta^{(v)})$ for views $v = 1, 2$. For each unlabeled instance $\mathbf{x} = [\mathbf{x}^{(1)}, \mathbf{x}^{(2)}]$, view 1 virtually splits it into two copies with opposite labels and fractional weights: $(\mathbf{x}, y = 1)$ with weight $P(y = 1|\mathbf{x}^{(1)}; \theta^{(1)})$ and $(\mathbf{x}, y = -1)$ with weight $1 - P(y = 1|\mathbf{x}^{(1)}; \theta^{(1)})$. View 1 then adds *all* augmented unlabeled instances to L_2. This is equivalent to the E-step in the EM algorithm. The same is true for view 2. Each view's parameter $\theta^{(v)}$ is then updated, which corresponds to the M-step, except that the expectations are from the other view. For certain tasks, Co-EM empirically performs better than Co-Training.

4.4   MULTIVIEW LEARNING*

The Co-Training algorithm is a means to an end: making the two classifiers $f^{(1)}$ and $f^{(2)}$ agree (i.e., predict the same label) on the unlabeled data. Such agreement is justified by learning theory, which is beyond the scope of this book, but the intuition is simple: there are not many candidate predictors that can agree on unlabeled data in two views, so the so-called hypothesis space is small. If a candidate predictor in this small hypothesis space also fits the labeled data well, it is less likely to be overfitting, and can be expected to be a good predictor. In this section we discuss some other algorithms which explicitly enforce hypothesis agreement, without requiring explicit feature splits or the iterative mutual-teaching procedure. To understand these algorithms, we need to introduce the regularized risk minimization framework for machine learning.

Recall that, in general, we can define a *loss function* to specify the cost of mistakes in prediction:

Definition 4.3. Loss Function. Let $\mathbf{x} \in \mathcal{X}$ be an instance, $y \in \mathcal{Y}$ its true label, and $f(\mathbf{x})$ our prediction. A loss function $c(\mathbf{x}, y, f(\mathbf{x})) \in [0, \infty)$ measures the amount of loss, or cost, of this prediction.

For example, in regression we can define the squared loss $c(\mathbf{x}, y, f(\mathbf{x})) = (y - f(\mathbf{x}))^2$. In classification we can define the 0/1 loss as $c(\mathbf{x}, y, f(\mathbf{x})) = 1$ if $y \neq f(\mathbf{x})$, and 0 otherwise. The loss function can be different for different types of misclassification. In medical diagnosis we might use $c(\mathbf{x}, y = \text{healthy}, f(\mathbf{x}) = \text{diseased}) = 1$ and $c(\mathbf{x}, y = \text{diseased}, f(\mathbf{x}) = \text{healthy}) = 100$. The loss function can also depend on the instance $\mathbf{x}$: The same amount of medical prediction error might incur a higher loss on an infant than on an adult.

Definition 4.4. Empirical Risk. The empirical risk of f is the average loss incurred by f on a labeled training sample: $\hat{R}(f) = \frac{1}{l} \sum_{i=1}^{l} c(\mathbf{x}_i, y_i, f(\mathbf{x}_i))$.

Applying the principle of *empirical risk minimization* (ERM)—finding the f that minimizes the empirical risk—may seem like a natural thing to do:

$$f^{ERM} = \operatorname*{argmin}_{f \in \mathcal{F}} \hat{R}(f), \tag{4.2}$$

where $\mathcal{F}$ is the set of all hypotheses we consider. For classification with 0/1 loss, ERM amounts to minimize the training sample error.

However, f^{ERM} can overfit the particular training sample. As a consequence, f^{ERM} is not necessarily the classifier in $\mathcal{F}$ with the smallest true risk on future data. One remedy is to *regularize* the empirical risk by a regularizer $\Omega(f)$. The regularizer $\Omega(f)$ is a non-negative functional, i.e., it takes a function f as input and outputs a non-negative real value. The value is such that if f is "smooth" or "simple" in some sense, $\Omega(f)$ will be close to zero; if f is too zigzagged (i.e., it overfits and attempts to pass through all labeled training instances), $\Omega(f)$ is large.

Definition 4.5. Regularized Risk. The regularized risk is the weighted sum of the empirical risk and the regularizer, with weight $\lambda \geq 0$: $\hat{R}(f) + \lambda \Omega(f)$. The principle of *regularized risk minimization* is to find the f that minimizes the regularized risk:

$$f^* = \operatorname*{argmin}_{f \in \mathcal{F}} \hat{R}(f) + \lambda \Omega(f). \tag{4.3}$$

The success of regularized risk minimization depends on the regularizer $\Omega(f)$. Different regularizers imply different assumptions of the task. For example, a commonly used regularizer for $f(\mathbf{x}) = \mathbf{w}^\top \mathbf{x}$ is $\Omega(f) = \frac{1}{2}\|\mathbf{w}\|^2$. This particular regularizer penalizes the squared norm of the parameters $\mathbf{w}$. It is helpful to view f as a point whose coordinates are determined by $\mathbf{w}$ in the parameter space. An equivalent form for the optimization problem in (4.3) is

$$
\begin{aligned}
\min_{f \in \mathcal{F}} \quad & \hat{R}(f) \\
\text{subject to} \quad & \Omega(f) \leq s,
\end{aligned}
$$

where s is determined by λ. It becomes clear that the regularizer constrains the radius of the ball in the parameter space. Within the ball, the f that best fits the training data is chosen. This controls the complexity of f, and prevents overfitting.

Importantly, for semi-supervised learning, one can often define the regularizer $\Omega(f)$ using the unlabeled data. For example,

$$
\Omega(f) = \Omega_{SL}(f) + \lambda' \Omega_{SSL}(f), \tag{4.4}
$$

where $\Omega_{SL}(f)$ is a supervised regularizer, and $\Omega_{SSL}(f)$ is a semi-supervised regularizer that depends on the unlabeled data. When $\Omega_{SSL}(f)$ indeed fits the task, such regularization can produce a better f^* than that produced by $\Omega_{SL}(f)$ alone. We will next show how to define $\Omega_{SSL}(f)$ to encourage agreement among multiple hypotheses, and discuss other forms of $\Omega_{SSL}(f)$ in later chapters.

We are now ready to introduce *multiview learning*. We assume the algorithm has access to k separate learners. It is possible, but not necessary, for each learner to use a subset of the features of an instance $\mathbf{x}$. This is the generalization of Co-Training to k views, hence the name multiview. Alternatively, the learners might be of different types (e.g., decision tree, neural network, etc.) but take the same features of $\mathbf{x}$ as input. This is similar to the so-called ensemble method. In either case, the goal is for the k learners to produce hypotheses $f_1^*, \ldots, f_k^*$ to minimize the following regularized risk:

$$
\begin{aligned}
(f_1^*, \ldots, f_k^*) \quad = \operatorname{argmin}_{f_1, \ldots, f_k} \quad & \sum_{v=1}^{k} \left(\sum_{i=1}^{l} c(\mathbf{x}_i, y_i, f_v(\mathbf{x}_i)) + \lambda_1 \Omega_{SL}(f_v) \right) \\
& + \lambda_2 \sum_{u,v=1}^{k} \sum_{i=l+1}^{l+u} c(\mathbf{x}_i, f_u(\mathbf{x}_i), f_v(\mathbf{x}_i)).
\end{aligned} \tag{4.5}
$$

The intuition is for each hypothesis to not only minimize its own empirical risk, but also agree with all the other hypotheses. The first part of the multiview regularized risk is simply the sum of individual (supervised) regularized risks. The second part defines a semi-supervised regularizer, which measures the disagreement of those k hypotheses on unlabeled instances:

$$
\Omega_{SSL}(f_1, \ldots, f_k) = \sum_{u,v=1}^{k} \sum_{i=l+1}^{l+u} c(\mathbf{x}_i, f_u(\mathbf{x}_i), f_v(\mathbf{x}_i)). \tag{4.6}
$$

The pairwise disagreement is defined as the loss on an unlabeled instance x_i when pretending $f_u(x_i)$ is the label and $f_v(x_i)$ is the prediction. Such disagreement is to be minimized. The final prediction for input $\mathbf{x}$ is the label least objected to by all the hypotheses:

$$y(\mathbf{x}) = \underset{y \in \mathcal{Y}}{\operatorname{argmin}} \sum_{v=1}^{k} c(\mathbf{x}, y, f_v^*(\mathbf{x})). \qquad (4.7)$$

Different c and Ω_{SL} lead to different instantiations of multiview learning. We give a concrete example below.

Example 4.6. Two-View Linear Ridge Regression Let each instance have two views $\mathbf{x} = [\mathbf{x}^{(1)}, \mathbf{x}^{(2)}]$. Consider two linear regression functions $f^{(1)}(\mathbf{x}) = \mathbf{w}^\top \mathbf{x}^{(1)}$, $f^{(2)}(\mathbf{x}) = \mathbf{v}^\top \mathbf{x}^{(2)}$. Let the loss function be $c(\mathbf{x}, y, f(\mathbf{x})) = (y - f(\mathbf{x}))^2$. Let the supervised regularizers be $\Omega_{SL}(f^{(1)}) = \|\mathbf{w}\|^2$, $\Omega_{SL}(f^{(2)}) = \|\mathbf{v}\|^2$. This particular form of regularization, i.e., penalizing the ℓ_2 norm of the parameter, is known as ridge regression. The regularized risk minimization problem is

$$\begin{aligned}
\min_{w,v} \quad & \sum_{i=1}^{l}(y_i - \mathbf{w}^\top \mathbf{x}_i^{(1)})^2 + \sum_{i=1}^{l}(y_i - \mathbf{v}^\top \mathbf{x}_i^{(2)})^2 + \lambda_1\|\mathbf{w}\|^2 + \lambda_1\|\mathbf{v}\|^2 \\
& +\lambda_2 \sum_{i=l+1}^{l+u}(\mathbf{w}^\top \mathbf{x}_i^{(1)} - \mathbf{v}^\top \mathbf{x}_i^{(2)})^2.
\end{aligned} \qquad (4.8)$$

The solution can be found by setting the gradient to zero and solving a system of linear equations.

What is the assumption behind multiview learning? In a regularized risk framework, the assumption is encoded in the regularizer Ω_{SSL} (4.6) to be minimized. That is, multiple hypotheses $f_1, \ldots, f_k$ should agree with each other. However, agreement alone is not sufficient. Consider the following counter-example: Replicate the feature k times to create k identical "views." Also replicate the hypotheses $f_1 = \ldots = f_k$. By definition they all agree, but this does not guarantee that they are any better than single-view learning (in fact the two are the same). The key insight is that the set of agreeing hypotheses need to additionally be a *small* subset of the hypothesis space $\mathcal{F}$. In contrast, the duplicating hypotheses in the counter-example still occupy the whole hypothesis space $\mathcal{F}$.

Remark 4.7. Multiview Learning Assumption Multiview learning is effective, when a set of hypotheses $f_1, \ldots, f_k$ agree with each other. Furthermore, there are not many such agreeing sets, and the agreeing set happens to have a small empirical risk.

This concludes our discussion of co-training and multiview learning techniques. These models use multiple views or classifiers, in conjunction with unlabeled data, in order to reduce the size of the hypothesis space. We also introduced the regularized risk minimization framework for machine learning, which will appear again in the next two chapters on graph-based methods and semi-supervised support vector machines.

BIBLIOGRAPHICAL NOTES

Co-Training was proposed by Blum and Mitchell [22, 129]. For simplicity, the algorithm presented here is slightly different from the original version. Further theoretical analysis of Co-Training can be found in [12, 10, 53]. Co-training has been applied to many tasks. For examples, see [41] and [93] on named entity classification in text processing. There are also many variants of co-training, including the Co-EM algorithm [134], single view [77, 38], single-view multiple-learner Democratic Co-learning algorithm [201], Tri-Training [206], Canonical Correlation Analysis [204] and relaxation of the conditional independence assumption [92].

Multiview learning was proposed as early as in [56]. It has been applied to semi-supervised regression [25, 159], and the more challenging problem of classification with structured outputs [24, 26]. Some theoretical analysis on the value of agreement among multiple learners can be found in [65, 110, 154, 193].

CHAPTER 5

Graph-Based Semi-Supervised Learning

5.1 UNLABELED DATA AS STEPPING STONES

Alice was flipping through the magazine "Sky and Earth," in which each article is either about astronomy or travel. Speaking no English, she had to guess the topic of each article from its pictures. The first story "Bright Asteroid" had a picture of a cratered asteroid—it was obviously about astronomy. The second story "Yellowstone Camping" had a picture of grizzly bears—she figured it must be a travel article.

But no other articles had pictures. "What is the use of a magazine without pictures?" thought Alice. The third article was titled "Zodiac Light," while the fourth "Airport Bike Rental." Not knowing any words and without pictures, it seemed impossible to guess the topic of these articles.

However, Alice is a resourceful person. She noticed the titles of other articles include "Asteroid and Comet," "Comet Light Curve," "Camping in Denali," and "Denali Airport." "I'll assume that if two titles share a word, they are about the same topic," she thought. And she started to doodle:

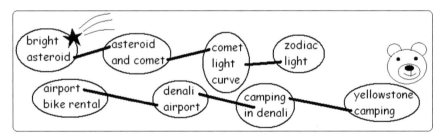

Alice's doodle. Articles sharing title words are connected.

Then it became clear. "Aha! 'Zodiac Light' is about astronomy, and 'Airport Bike Rental' is about travel!" exclaimed Alice. And she was correct. Alice just performed graph-based semi-supervised learning without knowing it.

5.2 THE GRAPH

Graph-based semi-supervised learning starts by constructing a graph from the training data. Given training data $\{(\mathbf{x}_i, y_i)\}_{i=1}^{l}$, $\{\mathbf{x}_j\}_{j=l+1}^{l+u}$, the vertices are the labeled and unlabeled instances $\{(\mathbf{x}_i)\}_{i=1}^{l} \cup \{\mathbf{x}_j\}_{j=l+1}^{l+u}$. Clearly, this is a large graph if u, the unlabeled data size, is big. Note that once the graph is built, learning will involve assigning y values to the vertices in the graph. This is made possible by edges that connect labeled vertices to unlabeled vertices. The graph edges are usually undirected. An edge between two vertices $\mathbf{x}_i, \mathbf{x}_j$ represents the similarity of the two instances. Let w_{ij} be the edge weight. The idea is that if w_{ij} is large, then the two labels y_i, y_j are expected to be the same. Therefore, the graph edge weights are of great importance. People often specify the edge weights with one of the following heuristics:

- Fully connected graph, where every pair of vertices $\mathbf{x}_i, \mathbf{x}_j$ is connected by an edge. The edge weight decreases as the Euclidean distance $\|\mathbf{x}_i - \mathbf{x}_j\|$ increases. One popular weight function is

$$ w_{ij} = \exp\left(-\frac{\|\mathbf{x}_i - \mathbf{x}_j\|^2}{2\sigma^2} \right), \tag{5.1} $$

  where σ is known as the bandwidth parameter and controls how quickly the weight decreases. This weight has the same form as a Gaussian function. It is also called a Gaussian kernel or a Radial Basis Function (RBF) kernel. The weight is 1 when $\mathbf{x}_i = \mathbf{x}_j$, and 0 when $\|\mathbf{x}_i - \mathbf{x}_j\|$ approaches infinity.

- kNN graph. Each vertex defines its k nearest neighbor vertices in Euclidean distance. Note if $\mathbf{x}_i$ is among $\mathbf{x}_j$'s kNN, the reverse is not necessarily true: $\mathbf{x}_j$ may not be among $\mathbf{x}_i$'s kNN. We connect $\mathbf{x}_i, \mathbf{x}_j$ if one of them is among the other's kNN. This means that a vertex may have more than k edges. If $\mathbf{x}_i, \mathbf{x}_j$ are connected, the edge weight w_{ij} is either the constant 1, in which case the graph is said to be unweighted, or a function of the distance as in (5.1). If $\mathbf{x}_i, \mathbf{x}_j$ are not connected, $w_{ij} = 0$. kNN graph automatically adapts to the density of instances in feature space: in a dense region, the kNN neighborhood radius will be small; in a sparse region, the radius will be large. Empirically, kNN graphs with small k tends to perform well.

- ϵNN graph. We connect $\mathbf{x}_i, \mathbf{x}_j$ if $\|\mathbf{x}_i - \mathbf{x}_j\| \leq \epsilon$. The edges can either be unweighted or weighted. If $\mathbf{x}_i, \mathbf{x}_j$ are not connected, $w_{ij} = 0$. ϵNN graphs are easier to construct than kNN graphs.

These are very generic methods. Of course, better graphs can be constructed if one has knowledge of the problem domain, and can define better distance functions, connectivity, and edge weights.

Figure 5.1 shows an example graph, where the edges are sparse. Let $\mathbf{x}_1, \mathbf{x}_2$ be the two labeled instances (vertices). Recall that the edges represent the "same label" assumption. For an unlabeled instance $\mathbf{x}_3$, its label y_3 is assumed to be similar to its neighbors in the graph, which in turn are similar to the neighbor's neighbors. Through this sequence of unlabeled data stepping stones, y_3 is assumed to be more similar to y_1 than to y_2. This is significant because $\mathbf{x}_3$ is in fact closer to $\mathbf{x}_2$ in Euclidean distance; without the graph, one would assume y_3 is more similar to y_2.

Formally, this intuition corresponds to estimating a label function f on the graph so that it satisfies two things: (1) the prediction $f(\mathbf{x})$ is close to the given label y on labeled vertices; 2) f should be smooth on the whole graph. This can be expressed in a regularization framework, where the former is encoded by the loss function, and the latter is encoded by a special graph-based

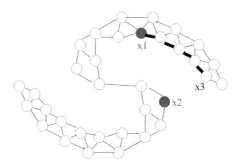

Figure 5.1: A graph constructed from labeled instances $\mathbf{x}_1, \mathbf{x}_2$ and unlabeled instances. The label of unlabeled instance $\mathbf{x}_3$ will be affected more by the label of $\mathbf{x}_1$, which is closer in the graph, than by the label of $\mathbf{x}_2$, which is farther in the graph, even though $\mathbf{x}_2$ is closer in Euclidean distance.

regularization. In the following sections, we introduce several different graph-based semi-supervised learning algorithms. They differ in the choice of the loss function and the regularizer. For simplicity, we will assume binary labels $y \in \{-1, 1\}$.

5.3 MINCUT

The first graph-based semi-supervised learning algorithm we introduce is formulated as a graph cut problem. We treat the positive labeled instances as "source" vertices, as if some fluid is flowing out of them and through the edges. Similarly, the negative labeled instances are "sink" vertices, where the fluid would disappear. The objective is to find a minimum set of edges whose removal blocks all flow from the sources to the sinks. This defines a "cut," or a partition of the graph into two sets of vertices. The "cut size" is measured by the sum of the weights on the edges defining the cut. Once the graph is split, the vertices connecting to the sources are labeled positive, and those to the sinks are labeled negative.

Mathematically, we want to find a function $f(\mathbf{x}) \in \{-1, 1\}$ on the vertices, such that $f(\mathbf{x}_i) = y_i$ for labeled instances, and the cut size is minimized:

$$\sum_{i, j: f(\mathbf{x}_i) \neq f(\mathbf{x}_j)} w_{ij}. \tag{5.2}$$

The above quantity is the cut size because if an edge w_{ij} is removed, it must be true that $f(\mathbf{x}_i) \neq f(\mathbf{x}_j)$.

We will now cast Mincut as a regularized risk minimization problem, with an appropriate loss function and regularizer. On any labeled vertex $\mathbf{x}_i$, $f(\mathbf{x}_i)$ is clamped at the given label: $f(\mathbf{x}_i) = y_i$. This can be enforced by a loss function of the form

$$c(\mathbf{x}, y, f(\mathbf{x})) = \infty \cdot (y - f(\mathbf{x}))^2, \tag{5.3}$$

where we define $\infty \cdot 0 = 0$. This loss function is zero if $f(\mathbf{x}_i) = y_i$, and infinity otherwise. To minimize the regularized risk, $f(\mathbf{x}_i)$ will then have to equal y_i on labeled vertices. The regularizer corresponds to the cut size. Recall we require $f(\mathbf{x}) \in \{-1, 1\}$ for all unlabeled vertices $\mathbf{x}$. Therefore, the cut size can be re-written as

$$\Omega(f) = \sum_{i,j=1}^{l+u} w_{ij}(f(\mathbf{x}_i) - f(\mathbf{x}_j))^2/4, \qquad (5.4)$$

Note the sum is now over *all* pairs of vertices. If $\mathbf{x}_i$ and $\mathbf{x}_j$ are not connected, then $w_{ij} = 0$ by definition; if the edge exists and is not cut, then $f(\mathbf{x}_i) - f(\mathbf{x}_j) = 0$. Thus, the cut size is well-defined even when we sum over all vertex pairs. In (5.4) we could have equivalently used $|f(\mathbf{x}_i) - f(\mathbf{x}_j)|/2$, but the squared term is consistent with other approaches discussed later in this chapter. The Mincut regularized risk problem is then

$$\min_{f:f(\mathbf{x})\in\{-1,1\}} \infty \sum_{i=1}^{l}(y_i - f(\mathbf{x}_i))^2 + \sum_{i,j=1}^{l+u} w_{ij}(f(\mathbf{x}_i) - f(\mathbf{x}_j))^2, \qquad (5.5)$$

where we scaled up both terms to remove the $1/4$. This is an integer programming problem because f is constrained to produce discrete values -1 or 1. However, efficient polynomial-time algorithms exist to solve the Mincut problem. It is clear that Mincut is a transductive learning algorithm, because the solution f is defined only on the vertices, not on the ambient feature space.

The formulation of Mincut has a curious "flaw" in that there could be multiple equally good solutions. For example, Figure 5.2 shows a graph with the shape of a chain. There are two labeled vertices: a positive vertex on one end, and a negative vertex on the other end. The edges have the same weight. There are six Mincut solutions: removing any single edge separates the two labels, and the cut size is minimized. What is wrong with six solutions? The label of the middle vertex is positive in three of the solutions, and negative in the other three. This label variability exists for other unlabeled vertices to a lesser extent. Intuitively, it seems to reflect the *confidence* of the labels. As we will see next, there are better ways to compute such confidence.

Figure 5.2: On this unweighted chain graph with one labeled vertex on each end, any single-edge cut is an equally good solution for Mincut.

5.4   HARMONIC FUNCTION

The second graph-based semi-supervised learning algorithm we introduce is the harmonic function. In our context, a harmonic function is a function that has the same values as given labels on the labeled data, and satisfies the weighted average property on the unlabeled data:

$$
\begin{aligned}
f(\mathbf{x}_i) &= y_i, \quad i = 1 \ldots l \\
f(\mathbf{x}_j) &= \frac{\sum_{k=1}^{l+u} w_{jk} f(\mathbf{x}_k)}{\sum_{k=1}^{l+u} w_{jk}}, \quad j = l+1 \ldots l+u.
\end{aligned}
\tag{5.6}
$$

In other words, the value assigned to each unlabeled vertex is the weighted average of its neighbors' values. The harmonic function is the solution to the same problem in (5.5), except that we relax f to produce real values:

$$
\min_{f:f(\mathbf{x}) \in \mathbb{R}} \infty \sum_{i=1}^{l} (y_i - f(\mathbf{x}_i))^2 + \sum_{i,j=1}^{l+u} w_{ij}(f(\mathbf{x}_i) - f(\mathbf{x}_j))^2.
\tag{5.7}
$$

This is equivalent to the more natural problem

$$
\begin{aligned}
\min_{f:f(\mathbf{x}) \in \mathbb{R}} \quad & \sum_{i,j=1}^{l+u} w_{ij}(f(\mathbf{x}_i) - f(\mathbf{x}_j))^2 \\
\text{subject to} \quad & f(\mathbf{x}_i) = y_i \text{ for } i = 1 \ldots l.
\end{aligned}
\tag{5.8}
$$

The relaxation has a profound effect. Now there is a closed-form solution for f. The solution is unique (under mild conditions) and globally optimal. The drawback of the relaxation is that in the solution, $f(\mathbf{x})$ is now a real value in $[-1, 1]$ that does not directly correspond to a label. This can however be addressed by thresholding $f(\mathbf{x})$ at zero to produce discrete label predictions (i.e., if $f(\mathbf{x}) >= 0$, predict $y = 1$, and if $f(\mathbf{x}) < 0$, predict $y = -1$).

The harmonic function f has many interesting interpretations. For example, one can view the graph as an electric network. Each edge is a resistor with resistance $1/w_{ij}$, or equivalently conductance w_{ij}. The labeled vertices are connected to a 1-volt battery, so that the positive vertices connect to the positive side, and the negative vertices connect to the ground. Then the voltage established at each node is the harmonic function,[1] see Figure 5.3(a).

The harmonic function f can also be interpreted by a random walk on the graph. Imagine a particle at vertex i. In the next time step, the particle will randomly move to another vertex j with probability proportional to w_{ij}:

$$
P(j|i) = \frac{w_{ij}}{\sum_k w_{ik}}.
\tag{5.9}
$$

The random walk continues in this fashion until the particle reaches one of the labeled vertices. This is known as an absorbing random walk, where the labeled vertices are absorbing states. Then the

---

[1]This, and the random walk interpretation below, is true when the labels $y \in \{0, 1\}$. When the labels $y \in \{-1, 1\}$, the voltages correspond to a shifted and scaled harmonic function.

value of the harmonic function at vertex i, $f(\mathbf{x}_i)$, is the probability that a particle starting at vertex i eventually reaches a positive labeled vertex (see Figure 5.3(b)).

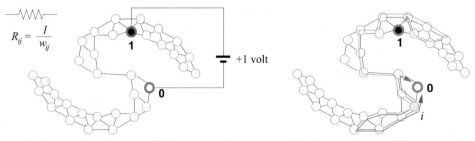

(a) The electric network interpretation   (b) The random walk interpretation

Figure 5.3: The harmonic function can be interpreted as the voltages of an electric network, or the probability of reaching a positive vertex in an absorbing random walk on the graph.

There is an iterative procedure to compute the harmonic function in (5.7). Initially, set $f(\mathbf{x}_i) = y_i$ for the labeled vertices $i = 1 \ldots l$, and some arbitrary value for the unlabeled vertices. Iteratively update each unlabeled vertex's f value with the weighted average of its neighbors:

$$f(\mathbf{x}_i) \leftarrow \frac{\sum_{j=1}^{l+u} w_{ij} f(\mathbf{x}_j)}{\sum_{j=1}^{l+u} w_{ij}}. \tag{5.10}$$

It can be shown that this iterative procedure is guaranteed to converge to the harmonic function, regardless of the initial values on the unlabeled vertices. This procedure is sometimes called label propagation, as it "propagates" labels from the labeled vertices (which are fixed) gradually through the edges to all the unlabeled vertices.

Finally, let us discuss the closed-form solution for the harmonic function. The solution is easier to present if we introduce some matrix notation. Let W be an $(l + u) \times (l + u)$ weight matrix, whose i, j-th element is the edge weight w_{ij}. Because the graph is undirected, W is a symmetric matrix. Its elements are non-negative. Let $D_{ii} = \sum_{j=1}^{l+u} w_{ij}$ be the weighted degree of vertex i, i.e., the sum of edge weights connected to i. Let D be the $(l + u) \times (l + u)$ diagonal matrix by placing $D_{ii}, i = 1 \ldots l + u$ on the diagonal. The unnormalized graph Laplacian matrix L is defined as

$$L = D - W. \tag{5.11}$$

Let $\mathbf{f} = (f(\mathbf{x}_1), \ldots, f(\mathbf{x}_{l+u}))^\top$ be the vector of f values on all vertices. The regularizer in (5.7) can be written as

$$\frac{1}{2} \sum_{i,j=1}^{l+u} w_{ij}(f(\mathbf{x}_i) - f(\mathbf{x}_j))^2 = \mathbf{f}^\top L \mathbf{f}. \tag{5.12}$$

Recall L is an $(l + u) \times (l + u)$ matrix. Assuming the vertices are ordered so that the labeled ones are listed first, we can partition the Laplacian matrix into four sub-matrices

$$L = \begin{bmatrix} L_{ll} & L_{lu} \\ L_{ul} & L_{uu} \end{bmatrix}, \tag{5.13}$$

partition $\mathbf{f}$ into $(\mathbf{f}_l, \mathbf{f}_u)$, and let $\mathbf{y}_l = (y_1, \ldots, y_l)^\top$. Then solving the constrained optimization problem using Lagrange multipliers with matrix algebra, one can show that the harmonic solution is

$$\begin{aligned} \mathbf{f}_l &= \mathbf{y}_l \\ \mathbf{f}_u &= -L_{uu}^{-1} L_{ul} \mathbf{y}_l. \end{aligned} \tag{5.14}$$

Example 5.1. Harmonic Function on Chain Graph Consider the chain graph in Figure 5.2. With the natural left-to-right order of vertices, we have

$$W = \begin{bmatrix} 0 & 1 & 0 & 0 & 0 & 0 & 0 \\ 1 & 0 & 1 & 0 & 0 & 0 & 0 \\ 0 & 1 & 0 & 1 & 0 & 0 & 0 \\ 0 & 0 & 1 & 0 & 1 & 0 & 0 \\ 0 & 0 & 0 & 1 & 0 & 1 & 0 \\ 0 & 0 & 0 & 0 & 1 & 0 & 1 \\ 0 & 0 & 0 & 0 & 0 & 1 & 0 \end{bmatrix} \quad D = \begin{bmatrix} 1 & 0 & 0 & 0 & 0 & 0 & 0 \\ 0 & 2 & 0 & 0 & 0 & 0 & 0 \\ 0 & 0 & 2 & 0 & 0 & 0 & 0 \\ 0 & 0 & 0 & 2 & 0 & 0 & 0 \\ 0 & 0 & 0 & 0 & 2 & 0 & 0 \\ 0 & 0 & 0 & 0 & 0 & 2 & 0 \\ 0 & 0 & 0 & 0 & 0 & 0 & 1 \end{bmatrix}$$

$$L = \begin{bmatrix} 1 & -1 & 0 & 0 & 0 & 0 & 0 \\ -1 & 2 & -1 & 0 & 0 & 0 & 0 \\ 0 & -1 & 2 & -1 & 0 & 0 & 0 \\ 0 & 0 & -1 & 2 & -1 & 0 & 0 \\ 0 & 0 & 0 & -1 & 2 & -1 & 0 \\ 0 & 0 & 0 & 0 & -1 & 2 & -1 \\ 0 & 0 & 0 & 0 & 0 & -1 & 1 \end{bmatrix} \tag{5.15}$$

To apply (5.14), we need to permute the order of vertices to be $(1, 7, 2, 3, 4, 5, 6)$ so that the labeled vertices come first. Also note $\mathbf{y}_l = (1, -1)^\top$. This gives

$$\mathbf{f}_u = \left(\frac{2}{3}, \frac{1}{3}, 0, -\frac{1}{3}, -\frac{2}{3}\right)^\top, \tag{5.16}$$

for the unlabeled vertices from left to right. It can be thresholded at zero to produce binary labels. This solution fits our intuition. The magnitude of the solution also coincides with label confidence.

5.5   MANIFOLD REGULARIZATION*

Both Mincut and the harmonic function are transductive learning algorithms. They each learn a function f that is restricted to the labeled and unlabeled vertices in the graph. There is no direct way to predict the label on a test instance $\mathbf{x}^*$ not seen before, unless one includes $\mathbf{x}^*$ as a new vertex into the graph and repeats the computation. This is clearly undesirable if we want predictions on a large number of test instances. What we need is an inductive semi-supervised learning algorithm.

In addition, both Mincut and harmonic function fix $f(\mathbf{x}) = y$ for labeled instances. What if some of the labels are wrong? It is not uncommon for real datasets to have such label noise. We want f to be able to occasionally disagree with the given labels.

Manifold regularization addresses these two issues. It is an inductive learning algorithm by defining f in the whole feature space: $f : \mathcal{X} \mapsto \mathbb{R}$. f is regularized to be smooth with respect to the graph by the graph Laplacian as in (5.12). However, this regularizer alone only controls $\mathbf{f}$, the value of f on the $l + u$ training instances. To prevent f from being too wiggly (and thus having inferior generalization performance) outside the training samples, it is necessary to include a second regularization term, such as $\|f\|^2 = \int_{x \in \mathcal{X}} f(x)^2 dx$. Putting them together, the regularizer for manifold regularization becomes

$$\Omega(f) = \lambda_1 \|f\|^2 + \lambda_2 \mathbf{f}^\top L \mathbf{f}, \tag{5.17}$$

where $\lambda_1, \lambda_2 \geq 0$ control the relative strength of the two terms. To allow f to disagree with the given labels, we can simply use the loss function $c(\mathbf{x}, y, f(\mathbf{x})) = (y - f(\mathbf{x}))^2$. This loss function does not greatly penalize small deviations. Other loss functions are possible, too, for example the hinge loss that we will introduce in Chapter 6. The complete manifold regularization problem is

$$\min_{f : \mathcal{X} \mapsto \mathbb{R}} \sum_{i=1}^{l} (y_i - f(\mathbf{x}_i))^2 + \lambda_1 \|f\|^2 + \lambda_2 \mathbf{f}^\top L \mathbf{f}. \tag{5.18}$$

The so-called representer theorem guarantees that the optimal f admits a finite ($l + u$, to be exact) dimensional representation. There exist efficient algorithms to find the optimal f.

Beyond the unnormalized graph Laplacian matrix L, the normalized graph Laplacian matrix $\mathcal{L}$ is often used too:

$$\mathcal{L} = D^{-1/2} L D^{-1/2} = I - D^{-1/2} W D^{-1/2}. \tag{5.19}$$

This results in a slightly different regularization term

$$\mathbf{f}^\top \mathcal{L} \mathbf{f} = \frac{1}{2} \sum_{i,j=1}^{l+u} w_{ij} \left(\frac{f(\mathbf{x}_i)}{\sqrt{D_{ii}}} - \frac{f(\mathbf{x}_j)}{\sqrt{D_{jj}}} \right)^2. \tag{5.20}$$

Other variations like L^p or $\mathcal{L}^p$, where $p > 0$, are possible too. They replace the matrix L in (5.18). These all encode the same overall label-smoothness assumption on the graph, but with varying subtleties. We discuss several properties of L below. Please see the references at the end of the chapter to learn more about the other variations.

5.6   THE ASSUMPTION OF GRAPH-BASED METHODS*

Remark 5.2. Graph-Based Semi-Supervised Learning Assumption   The labels are "smooth" with respect to the graph, such that they vary slowly on the graph. That is, if two instances are connected by a strong edge, their labels tend to be the same.

The notion of smoothness can be made precise by spectral graph theory. A vector ϕ is an eigenvector of a square matrix A, if

$$A\phi = \lambda\phi, \tag{5.21}$$

where λ is the associated eigenvalue. If ϕ is an eigenvector, $c\phi$ is an eigenvector, too, for any $c \neq 0$. But we will focus on eigenvectors of unit length $\|\phi\| = 1$. Spectral graph theory is concerned with the eigenvectors and eigenvalues of a graph, represented by its Laplacian matrix L or $\mathcal{L}$. We will analyze the unnormalized Laplacian L, which has the following properties:

- L has $l + u$ eigenvalues (some may be the same) and corresponding eigenvectors $\{(\lambda_i, \phi_i)\}_{i=1}^{l+u}$. These pairs are called the graph spectrum. The eigenvectors are orthogonal: $\phi_i^\top \phi_j = 0$ for $i \neq j$.

- The Laplacian matrix can be decomposed into a weighted sum of outer products:

$$L = \sum_{i=1}^{l+u} \lambda_i \phi_i \phi_i^\top. \tag{5.22}$$

- The eigenvalues are non-negative real numbers, and can be sorted as

$$0 = \lambda_1 \leq \lambda_2 \leq \ldots \leq \lambda_{l+u}. \tag{5.23}$$

In particular, the graph has k connected components if and only if $\lambda_1 = \ldots = \lambda_k = 0$. The corresponding eigenvectors are constant on individual connected components, and zero elsewhere, as the following example shows.

Example 5.3. The Spectrum of a Disconnected Chain Graph   Figure 5.4(a) shows a disconnected, unweighted chain graph with 20 vertices. Its spectrum is shown in Figure 5.4(b). The stem plots are the corresponding eigenvectors ϕ_i. Note $\lambda_1 = \lambda_2 = 0$ since there are two connected components. The eigenvectors ϕ_1, ϕ_2 are piecewise constant, whose height is determined by length normalization. As the eigenvalues increase in magnitude, the corresponding eigenvectors become more and more rugged.

Because the eigenvectors are orthogonal and have unit length, they form a basis in $\mathbb{R}^{l+u}$. This means any $\mathbf{f}$ on the graph can be decomposed into

$$\mathbf{f} = \sum_{i=1}^{l+u} a_i \phi_i, \tag{5.24}$$

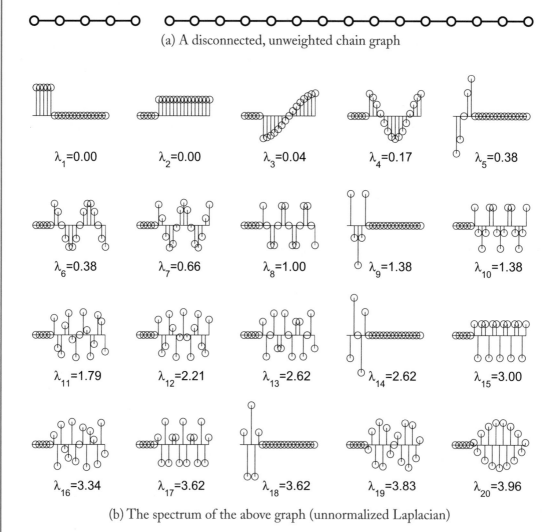

(a) A disconnected, unweighted chain graph

$\lambda_1=0.00$ $\lambda_2=0.00$ $\lambda_3=0.04$ $\lambda_4=0.17$ $\lambda_5=0.38$

$\lambda_6=0.38$ $\lambda_7=0.66$ $\lambda_8=1.00$ $\lambda_9=1.38$ $\lambda_{10}=1.38$

$\lambda_{11}=1.79$ $\lambda_{12}=2.21$ $\lambda_{13}=2.62$ $\lambda_{14}=2.62$ $\lambda_{15}=3.00$

$\lambda_{16}=3.34$ $\lambda_{17}=3.62$ $\lambda_{18}=3.62$ $\lambda_{19}=3.83$ $\lambda_{20}=3.96$

(b) The spectrum of the above graph (unnormalized Laplacian)

Figure 5.4: The spectrum of a disconnected chain graph, showing two connected components, and the smooth to rough transition of the eigenvectors.

where $a_i, i = 1 \ldots l + u$ are real-valued coefficients. It is possible to show that the graph regularizer (5.12) can be written as

$$\mathbf{f}^\top L \mathbf{f} = \sum_{i=1}^{l+u} a_i^2 \lambda_i. \qquad (5.25)$$

If for each i, either a_i or λ_i is close to zero, then $\mathbf{f}^\top L \mathbf{f}$ will be small. Intuitively, this means the graph regularizer $\mathbf{f}^\top L \mathbf{f}$ prefers an $\mathbf{f}$ that only uses smooth basis (those with a small λ_i). By "uses" we mean $\mathbf{f}$'s corresponding coefficients have large magnitude $|a_i|$. In particular, $\mathbf{f}^\top L \mathbf{f}$ is minimized and equals zero, if $\mathbf{f}$ is in the subspace spanned by $\phi_1, \ldots, \phi_k$ for a graph with k connected components:

$$\mathbf{f} = \sum_{i=1}^{k} a_i \phi_i, \quad a_i = 0 \text{ for } i > k. \qquad (5.26)$$

For a connected graph, only $\lambda_1 = 0$, and $\phi_1 = (1/\sqrt{l+u}, \ldots, 1/\sqrt{l+u})$. Any constant vector $\mathbf{f}$ thus has coefficients $a_1 \neq 0$, $a_i = 0$ for $i > 1$, and is a minimizer of $\mathbf{f}^\top L \mathbf{f}$. Being a constant, it is certainly the most smooth function on the graph.

Therefore, we see the connection between graph-based semi-supervised learning methods and the graph spectrum. This exposes a major weakness of this family of methods: the performance is sensitive to the graph structure and edge weights.

Example 5.4. Harmonic Function with a Bad Graph   To demonstrate the previous point, Figure 5.5 presents a dataset comprised of two semi-circles (one per class) that intersect. This creates a problem for the types of graphs traditionally used in graph-based semi-supervised learning. Figure 5.5(a) shows the symmetric 4-NN graph for this data. An ϵNN graph is similar. Note that, near the intersection of the two curves, many edges connect instances on one curve to instances on the other. This means that any algorithm that assumes label smoothness with respect to the graph is likely to make many mistakes in this region. Labels will propagate between the classes, which is clearly undesirable.

We applied the harmonic function with only two labeled instances (large X and O) using the graph shown and weights as in (5.1). Figure 5.5(b) shows the predicted labels (small x's and o's) for the unlabeled instances. We observe that the predictions are very poor, as the O label propagates through most of the other class's instances due to the between-class connections in the graph. Even a simple linear classifier using only the labeled instances would be able to correctly predict the left half of the X curve (i.e., the decision boundary would be a diagonal line between the labeled instances). While graph-based semi-supervised learning can be a powerful method to incorporate unlabeled data, one must be careful to ensure that the graph encodes the correct label smoothness assumption.

To handle this dataset properly and obtain all correct predictions, the graph would need to split the data into two disconnected components. One approach to building such a graph is to examine the local neighborhood around each instance and only connect instances whose neighborhoods have similar shapes: neighborhoods along the *same* curve would look similar with only a minor rotation, while neighborhoods *across* curves would be rotated 90 degrees. The resulting graph should avoid inter-class edges. Results using a similar approach for several datasets like the one seen here can be found in [75].

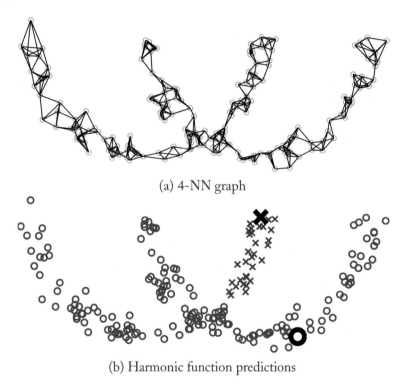

(a) 4-NN graph

(b) Harmonic function predictions

Figure 5.5: Graph-based semi-supervised learning using a bad graph can lead to poor performance.

This chapter introduced the notion of using a graph over labeled and unlabeled data to perform semi-supervised learning. We discussed several algorithms that share the intuition that the predictions should be smooth with respect to this graph. We introduced some notions from spectral graph theory to justify this approach, and illustrated what can go wrong if the graph is not constructed carefully. In the next chapter, we discuss semi-supervised support vector machines, which make a very different assumption about the space containing the data.

BIBLIOGRAPHICAL NOTES

The idea of the target function being smooth on the graph, or equivalently regularization by the graph, is very natural. Therefore, there are many related methods that exploit this idea, including Mincut [21] and randomized Mincut [20], Boltzmann machines [70, 209], graph random

walk [8, 168], harmonic function [210], local and global consistency [198], manifold regularization [17, 158, 155], kernels from the graph Laplacian [35, 51, 95, 101, 163, 211], spectral graph
transducer [90], local averaging [179, 187], density-based regularization [23, 36], alternating minimization [181], boosting [39, 117], and the tree-based Bayes model [98]. The graph construction
itself is important [11, 83, 82, 29, 167, 196]. Some theoretical analysis of graph-based learning can
be found in [91, 178, 195], and applications in [73, 78, 111, 102, 136, 138, 139].

Many of the graph-based semi-supervised learning algorithms have moderate to high computational complexity, often $O(u^2)$ or more. Fast computation to handle large amount of unlabeled
data is an important problem. Efforts toward this end include [6, 57, 69, 84, 123, 160, 172, 192, 212].
Alternatively, one can perform online semi-supervised learning [72] where the labeled and unlabeled
instances arrive sequentially. They are processed and discarded immediately to keep the computation
and storage requirement low.

Manifold regularization [17] formally assumes that the marginal distribution $p(\mathbf{x})$ is supported
on a Riemannian manifold (see Chapter 2 in [107] for a brief introduction). The labeled and unlabeled
vertices, and hence the graph, are a random realization of the underlying manifold. For simplicity,
we did not introduce this assumption during our discussion.

There are several extensions to the simple undirected graph that encodes similarity between
vertices. In certain applications like the Web, the edges naturally are directed [27, 118, 200]. Some
graph edges might encode dissimilarities instead [76, 171]. Edges can also be defined on more than
two vertices to form hypergraphs [199]. The dataset can consist of multiple manifolds [74, 180, 197].

CHAPTER 6

Semi-Supervised Support Vector Machines

The intuition behind Semi-Supervised Support Vector Machines (S3VMs) is very simple. Figure 6.1(a) shows a completely labeled dataset. If we were to draw a straight line to separate the two classes, where should the line be? One reasonable place is right in the middle, such that its distance to the nearest positive or negative instance is maximized. This is the linear decision boundary found by Support Vector Machines (SVMs), and is shown in Figure 6.1(a). The figure also shows two dotted lines that go through the nearest positive and negative instances. The distance from the decision boundary to a dotted line is called the geometric margin. As mentioned above, this margin is maximized by SVMs.

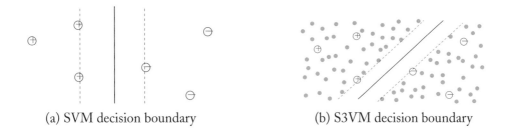

(a) SVM decision boundary   (b) S3VM decision boundary

Figure 6.1: (a) With only labeled data, the linear decision boundary that maximizes the distance to any labeled instance is shown in solid line. Its associated margin is shown in dashed lines. (b) With additional unlabeled data, under the assumption that the classes are well-separated, the decision boundary seeks a gap in unlabeled data.

What if we have many additional unlabeled instances, distributed as in Figure 6.1(b)? The SVM decision boundary will cut through dense unlabeled data regions. This seems undesirable, if we assume that the two classes are well-separated. Instead, the best decision boundary now seems to be the one in Figure 6.1(b), which falls in to the gap between the unlabeled data. This new decision boundary still separates the two classes in the labeled data, though its margin is smaller than the SVM decision boundary (this can be easily verified by measuring the distance to the nearest labeled point). The new decision boundary is the one found by S3VMs, and is defined by both labeled and unlabeled data. We will now make this intuition precise.

6.1   SUPPORT VECTOR MACHINES

We first discuss SVMs. Our discussion is intended to be just enough for the purpose of introducing S3VMs in the next section. For a complete exposition see any standard textbook, e.g., [49, 149, 176]. For simplicity, we will assume that there are two classes: $y \in \{-1, 1\}$. We will also assume that the decision boundary is linear in $\mathbb{R}^D$. Such a decision boundary is defined by the set

$$\{\mathbf{x} | \mathbf{w}^\top \mathbf{x} + b = 0\}, \tag{6.1}$$

where $\mathbf{w} \in \mathbb{R}^D$ is the parameter vector that specifies the orientation and scale of the decision boundary, and $b \in \mathbb{R}$ is an offset parameter. For example, when $\mathbf{w} = (1, 1)^\top$ and $b = -1$, the decision boundary is shown as the blue line in Figure 6.2. The decision boundary is always perpendicular to the vector $\mathbf{w}$. The value b determines the shift along $\mathbf{w}$.

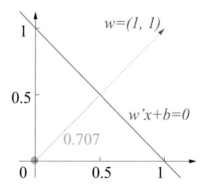

Figure 6.2: The linear decision boundary (the blue line) $\mathbf{w}^\top \mathbf{x} + b = 0$, where $\mathbf{w} = (1, 1)^\top$ (the red vector), and $b = -1$. The distance from point $(0, 0)$ to this decision boundary is $1/\sqrt{2}$ (the green line).

Let $f(\mathbf{x}) = \mathbf{w}^\top \mathbf{x} + b$. The decision boundary is thus defined by $f(\mathbf{x}) = 0$. We will predict the label of $\mathbf{x}$ by $\text{sign}(f(\mathbf{x}))$. We are interested in the distance between an instance $\mathbf{x}$ to the decision boundary. The absolute value of this distance turns out to be $|f(\mathbf{x})|/\|\mathbf{w}\|$. For example, the origin $\mathbf{x} = (0, 0)^\top$ has a distance $1/\sqrt{2} \approx 0.707$ to the decision boundary, as shown by the green line in Figure 6.2.

The decision boundary cuts the feature space into two halves, one half with $f > 0$ (the positive side), and the other half with $f < 0$ (the negative side). We define the *signed distance* of a labeled instance $(\mathbf{x}, y)$ to the decision boundary as

$$yf(\mathbf{x})/\|\mathbf{w}\|. \tag{6.2}$$

The signed distance is positive, if a positive instance is on the positive side, or a negative instance on the negative side. For now, we also assume that the training sample is *linearly separable*, meaning that there is at least one linear decision boundary that can separate all labeled instances so they are on the correct side of the decision boundary. The signed geometric margin is the distance from the decision boundary to the closest labeled instance:

$$\min_{i=1}^{l} y_i f(\mathbf{x}_i)/\|\mathbf{w}\|. \tag{6.3}$$

If a decision boundary separates the labeled training sample, the geometric margin is positive. We want to find the decision boundary that maximizes the geometric margin:

$$\max_{\mathbf{w},b} \min_{i=1}^{l} y_i f(\mathbf{x}_i)/\|\mathbf{w}\|. \tag{6.4}$$

This is difficult to optimize directly, so we will rewrite it into an equivalent form. First, we notice that one can arbitrarily scale the parameters $(\mathbf{w}, b) \to (c\mathbf{w}, cb)$ in (6.4). To remove this nuisance degree of freedom, we require that the instances closest to the decision boundary satisfy

$$\min_{i=1}^{l} y_i f(\mathbf{x}_i) = 1. \tag{6.5}$$

This implies that for all labeled instances $i = 1 \ldots l$, we have the constraint

$$y_i f(\mathbf{x}_i) = y_i(\mathbf{w}^\top \mathbf{x}_i + b) \geq 1. \tag{6.6}$$

We can then rewrite (6.4) as a constrained optimization problem:

$$\begin{aligned} \max_{\mathbf{w},b} \quad & 1/\|\mathbf{w}\| \\ \text{subject to} \quad & y_i(\mathbf{w}^\top \mathbf{x}_i + b) \geq 1, \quad i = 1 \ldots l. \end{aligned} \tag{6.7}$$

In addition, maximizing $1/\|\mathbf{w}\|$ is equivalent to minimizing $\|\mathbf{w}\|^2$. This leads to the following quadratic program, which is easier to optimize:

$$\begin{aligned} \min_{\mathbf{w},b} \quad & \|\mathbf{w}\|^2 \\ \text{subject to} \quad & y_i(\mathbf{w}^\top \mathbf{x}_i + b) \geq 1, \quad i = 1 \ldots l. \end{aligned} \tag{6.8}$$

So far, we have assumed that the training sample is linearly separable. This means the constraints in (6.8) can all be satisfied by at least *some* parameters $\mathbf{w}, b$. We now relax this assumption to allow any training sample, even linearly non-separable ones. When a training sample is linearly non-separable, at least one constraint in (6.8) cannot be satisfied by *any* parameters. This renders (6.8) infeasible. We have to relax the constraints by allowing $y_i f(\mathbf{x}_i) < 1$ on some instances. But we will penalize the amount by which we have to make this relaxation. This is done by introducing *slack variables*, i.e., the amount of relaxation for each instance, $\xi_i \geq 0, i = 1 \ldots l$ into (6.8):

$$\begin{aligned} \min_{\mathbf{w},b,\xi} \quad & \sum_{i=1}^{l} \xi_i + \lambda \|\mathbf{w}\|^2 \\ \text{subject to} \quad & y_i(\mathbf{w}^\top \mathbf{x}_i + b) \geq 1 - \xi_i, \quad i = 1 \ldots l \\ & \xi_i \geq 0. \end{aligned} \tag{6.9}$$

In the equation above, $\sum_{i=1}^{l} \xi_i$ is the total amount of relaxation, and we would like to minimize a weighted sum of it and $\|\mathbf{w}\|^2$. The weight λ balances the two objectives. This formulation thus still attempts to find the maximum margin separation, but allows some training instances to be on the wrong side of the decision boundary. It is still a quadratic program. The optimization problem (6.9) is known as the primal form of a linear SVM.

It is illustrative to cast (6.9) into a regularized risk minimization framework, as this is how we will extend it to S3VMs. Consider the following optimization problem:

$$
\begin{aligned}
\min_{\xi} \quad & \xi \\
\text{subject to} \quad & \xi \geq z \\
& \xi \geq 0.
\end{aligned}
\tag{6.10}
$$

It is easy to verify that when $z \leq 0$, the objective is 0; when $z > 0$, the objective is z. Therefore, solving problem (6.10) is equivalent to evaluating the function

$$
\max(z, 0).
\tag{6.11}
$$

Noting in (6.9) the inequality constraints on ξ_i can be written as $\xi_i \geq 1 - y_i(\mathbf{w}^\top \mathbf{x}_i + b)$, we set $z_i = 1 - y_i(\mathbf{w}^\top \mathbf{x}_i + b)$ to turn (6.9) into the sum of the form (6.10). This in turn converts (6.9) into the following equivalent, but unconstrained, regularized risk minimization problem

$$
\min_{\mathbf{w}, b} \sum_{i=1}^{l} \max(1 - y_i(\mathbf{w}^\top \mathbf{x}_i + b), 0) + \lambda \|\mathbf{w}\|^2,
\tag{6.12}
$$

where the first term corresponds to the loss function

$$
c(\mathbf{x}, y, f(\mathbf{x})) = \max(1 - y(\mathbf{w}^\top \mathbf{x} + b), 0),
\tag{6.13}
$$

and the second term corresponds to the regularizer

$$
\Omega(f) = \|\mathbf{w}\|^2.
\tag{6.14}
$$

The particular loss function (6.13) is known as the *hinge loss*. We plot hinge loss as a function of $yf(\mathbf{x}) = y(\mathbf{w}^\top \mathbf{x} + b)$ in Figure 6.3(a). Recall that for well-separated training instances, we have $yf(\mathbf{x}) \geq 1$. Therefore, the hinge loss penalizes instances which are on the correct side of the decision boundary, but within the margin ($0 \leq yf(\mathbf{x}) < 1$); it penalizes instances even more if they are on the wrong side of the decision boundary ($yf(\mathbf{x}) < 0$). The shape of the loss function resembles a hinge, hence the name.

We will not discuss the dual form of SVMs, nor the kernel trick that essentially maps the feature to a higher dimensional space to handle non-linear problems. These are crucial to the success of SVMs, but are not necessary to introduce S3VMs. However, we shall point out that it is straightforward to apply the kernel trick to S3VMs, too.

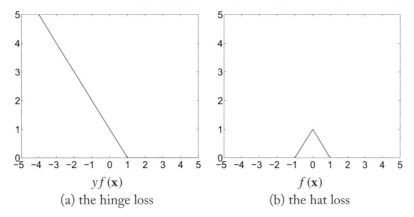

(a) the hinge loss

(b) the hat loss

Figure 6.3: (a) The hinge loss $c(\mathbf{x}, y, f(\mathbf{x})) = \max(1 - y(\mathbf{w}^\top \mathbf{x} + b), 0)$ as a function of $yf(\mathbf{x})$. (b) The hat loss $c(\mathbf{x}, \hat{y}, f(\mathbf{x})) = \max(1 - |\mathbf{w}^\top \mathbf{x} + b|, 0)$ as a function of $f(\mathbf{x})$.

6.2   SEMI-SUPERVISED SUPPORT VECTOR MACHINES*

Semi-Supervised Support Vector Machines (S3VMs) were originally called Transductive Support Vector Machines (TSVMs), because its theory was developed to give performance bounds (theoretical guarantees) on the given unlabeled sample. However, since the learned function f naturally applies to unseen test instances, it is more appropriate to call them S3VMs.

Recall that in Figure 6.1(b), the intuition of S3VM is to place *both* labeled and unlabeled instances outside the margin. We have seen how this can be encouraged for the labeled instances using the hinge loss in Figure 6.3(a). But what about unlabeled instances? Without a label, we do not even know whether an unlabeled instance $\mathbf{x}$ is on the correct or the wrong side of the decision boundary.

Here is one way to incorporate the unlabeled instance $\mathbf{x}$ into learning. Recall the label prediction on $\mathbf{x}$ is $\hat{y} = \text{sign}(f(\mathbf{x}))$. If we treat this prediction as the *putative label* of $\mathbf{x}$, then we can apply the hinge loss function on $\mathbf{x}$:

$$
\begin{aligned}
c(\mathbf{x}, \hat{y}, f(\mathbf{x})) &= \max(1 - \hat{y}(\mathbf{w}^\top \mathbf{x} + b), 0) \\
&= \max(1 - \text{sign}(\mathbf{w}^\top \mathbf{x} + b)(\mathbf{w}^\top \mathbf{x} + b), 0) \\
&= \max(1 - |\mathbf{w}^\top \mathbf{x} + b|, 0),
\end{aligned}
\tag{6.15}
$$

where we used the fact $\text{sign}(z)z = |z|$. This new loss function is distinct from the hinge loss in that it does not need the real label y, but is instead completely determined by $f(\mathbf{x})$. The new loss function is plotted in Figure 6.3(b). Note the x-axis is now $f(\mathbf{x})$ instead of $yf(\mathbf{x})$. Due to its distinctive shape, this new loss function in (6.15) is called the *hat loss*.

Because of the way we generate the putative label $\hat{y}$, an unlabeled instance $\mathbf{x}$ is *always* on the correct side of the decision boundary. Nonetheless, the hat loss still penalizes certain unlabeled instances. Specifically, it prefers $f(\mathbf{x}) \geq 1$ or $f(\mathbf{x}) \leq -1$ (the "rim" of the hat). These are unlabeled instances outside the margin, far away from the decision boundary. On the other hand, it penalizes unlabeled instances with $-1 < f(\mathbf{x}) < 1$, especially the ones with $f(\mathbf{x}) \approx 0$. These are unlabeled instances within the margin. Intuitively, they are the ones that f is uncertain about. We now incorporate the hat loss on the unlabeled data $\{\mathbf{x}_j\}_{j=l+1}^{l+u}$ into the SVM objective (6.12) to form the S3VM objective:

$$\min_{\mathbf{w},b} \sum_{i=1}^{l} \max(1 - y_i(\mathbf{w}^\top \mathbf{x}_i + b), 0) + \lambda_1 \|\mathbf{w}\|^2 + \lambda_2 \sum_{j=l+1}^{l+u} \max(1 - |\mathbf{w}^\top \mathbf{x}_j + b|, 0). \tag{6.16}$$

The S3VM objective prefers unlabeled instances to be outside the margin. Equivalently, the decision boundary would want to be in a low density gap in the dataset, such that few unlabeled instances are close. Although we used the name "hat loss," it is more natural to view (6.16) as regularized risk minimization with hinge loss on labeled instances, and a regularizer involving these hat-shaped functions:

$$\Omega(f) = \lambda_1 \|\mathbf{w}\|^2 + \lambda_2 \sum_{j=l+1}^{l+u} \max(1 - |\mathbf{w}^\top \mathbf{x}_j + b|, 0). \tag{6.17}$$

There is one practical consideration. Empirically, it is sometimes observed that the solution to (6.16) is imbalanced. That is, the majority (or even all) of the unlabeled instances are predicted in only one of the classes. The reason for such behavior is not well-understood. To correct for the imbalance, one heuristic is to constrain the predicted class proportion on the unlabeled data, so that it is the same as the class proportion on the labeled data:

$$\frac{1}{u} \sum_{j=l+1}^{l+u} \hat{y}_j = \frac{1}{l} \sum_{i=1}^{l} y_i. \tag{6.18}$$

Since $\hat{y}_j = \text{sign}(f(\mathbf{x}_j))$ is a discontinuous function, the constraint is difficult to enforce. Instead, we relax it into a constraint involving continuous functions:

$$\frac{1}{u} \sum_{j=l+1}^{l+u} f(\mathbf{x}_j) = \frac{1}{l} \sum_{i=1}^{l} y_i. \tag{6.19}$$

Therefore, the complete S3VM problem with class balance constraint is

$$\min_{\mathbf{w},b} \quad \sum_{i=1}^{l} \max(1 - y_i(\mathbf{w}^\top \mathbf{x}_i + b), 0) + \lambda_1 \|\mathbf{w}\|^2 + \lambda_2 \sum_{j=l+1}^{l+u} \max(1 - |\mathbf{w}^\top \mathbf{x}_j + b|, 0)$$

$$\text{subject to} \quad \frac{1}{u} \sum_{j=l+1}^{l+u} \mathbf{w}^\top \mathbf{x}_j + b = \frac{1}{l} \sum_{i=1}^{l} y_i. \tag{6.20}$$

Finally, we point out a computational difficulty of S3VMs. The S3VM objective function (6.16) is *non-convex*. A function g is convex, if for $\forall z_1, z_2, \forall 0 \leq \lambda \leq 1$,

$$g(\lambda z_1 + (1 - \lambda)z_2) \leq \lambda g(z_1) + (1 - \lambda)g(z_2). \tag{6.21}$$

For example, the SVM objective (6.12) is a convex function of the parameters $\mathbf{w}, b$. This can be verified by the convexity of the hinge loss, the squared norm, and the fact that the sum of convex functions is convex. Minimizing a convex function is relatively easy, as such a function has a well-defined "bottom." On the other hand, the hat loss function is non-convex, as demonstrated by $z_1 = -1, z_2 = 1$, and $\lambda = 0.5$. With the sum of a large number of hat functions, the S3VM objective (6.16) is non-convex with multiple local minima. A learning algorithm can get trapped in a sub-optimal local minimum, and not find the global minimum solution. The research in S3VMs has focused on how to efficiently find a near-optimum solution; some of this work is listed in the bibliographical notes.

6.3   ENTROPY REGULARIZATION*

SVMs and S3VMs are non-probabilistic models. That is, they are not designed to compute the label posterior probability $p(y|\mathbf{x})$ when making classification. In statistical machine learning, there are many probabilistic models which compute $p(y|\mathbf{x})$ from labeled training data for classification. Interestingly, there is a direct analogue of S3VM for these probabilistic models too, known as entropy regularization. To make our discussion concrete, we will first introduce a particular probabilistic model: logistic regression, and then extend it to semi-supervised learning via entropy regularization.

Logistic regression models the posterior probability $p(y|\mathbf{x})$. Like SVMs, it uses a linear decision function $f(\mathbf{x}) = \mathbf{w}^\top \mathbf{x} + b$. Let the label $y \in \{-1, 1\}$. Recall that if $f(\mathbf{x}) \gg 0, \mathbf{x}$ is deep within the positive side of the decision boundary, if $f(\mathbf{x}) \ll 0, \mathbf{x}$ is deep within the negative side, and $f(\mathbf{x}) = 0$ means $\mathbf{x}$ is right on the decision boundary with maximum label uncertainty. Logistic regression models the posterior probability by

$$p(y|\mathbf{x}) = 1/\left(1 + \exp(-y f(\mathbf{x}))\right), \tag{6.22}$$

which "squashes" $f(\mathbf{x}) \in (-\infty, \infty)$ down to $p(y|\mathbf{x}) \in [0, 1]$. The model parameters are $\mathbf{w}$ and b, like in SVMs. Given a labeled training sample $\{(\mathbf{x}_i, y_i)\}_{i=1}^l$, the conditional log likelihood is defined as

$$\sum_{i=1}^l \log p(y_i|\mathbf{x}_i, \mathbf{w}, b). \tag{6.23}$$

If we further introduce a Gaussian distribution as the prior on $\mathbf{w}$:

$$\mathbf{w} \sim \mathcal{N}(0, I/(2\lambda)) \tag{6.24}$$

where I is the diagonal matrix of the appropriate dimension, then logistic regression training is to maximize the posterior of the parameters:

$$\max_{\mathbf{w},b} \log p(\mathbf{w}, b | \{(\mathbf{x}_i, y_i)\}_{i=1}^{l})$$

$$= \max_{\mathbf{w},b} \log p(\{(\mathbf{x}_i, y_i)\}_{i=1}^{l} | \mathbf{w}, b) + \log p(\mathbf{w})$$

$$= \max_{\mathbf{w},b} \sum_{i=1}^{l} \log \left(1 / \left(1 + \exp(-y_i f(\mathbf{x}_i)) \right) \right) - \lambda \|\mathbf{w}\|^2. \tag{6.25}$$

The second line follows from Bayes rule, and ignoring the denominator that is constant with respect to the parameters. This is equivalent to the following regularized risk minimization problem:

$$\min_{\mathbf{w},b} \sum_{i=1}^{l} \log \left(1 + \exp(-y_i f(\mathbf{x}_i)) \right) + \lambda \|\mathbf{w}\|^2, \tag{6.26}$$

with the so-called logistic loss

$$c(\mathbf{x}, y, f(\mathbf{x})) = \log \left(1 + \exp(-y f(\mathbf{x})) \right), \tag{6.27}$$

and the usual regularizer $\Omega(f) = \|\mathbf{w}\|^2$. Figure 6.4(a) shows the logistic loss. Note its similarity to the hinge loss in Figure 6.3(a).

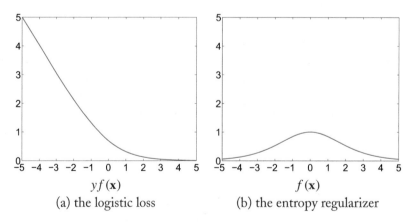

(a) the logistic loss   (b) the entropy regularizer

Figure 6.4: (a) The logistic loss $c(\mathbf{x}, y, f(\mathbf{x})) = \log \left(1 + \exp(-y f(\mathbf{x})) \right)$ as a function of $y f(\mathbf{x})$. (b) The entropy regularizer that encourages high-confidence classification on unlabeled data.

Logistic regression does not use unlabeled data. We can include unlabeled data based on the following intuition: if the two classes are well-separated, then the classification on any unlabeled instance should be confident: it either clearly belongs to the positive class, or to the negative class. Equivalently, the posterior probability $p(y|\mathbf{x})$ should be either close to 1, or close to 0. One way to measure the confidence is the entropy. For a Bernoulli random variable with probability p, the entropy is defined as

$$H(p) = -p \log p - (1-p) \log(1-p). \tag{6.28}$$

The entropy H reaches its minimum 0 when $p = 0$ or $p = 1$, i.e., when the outcome is most certain; H reaches its maximum 1 when $p = 0.5$, i.e., most uncertain. Given a unlabeled training sample $\{\mathbf{x}_j\}_{j=l+1}^{l+u}$, the entropy regularizer for logistic regression is defined as

$$\Omega(f) = \sum_{j=l+1}^{l+u} H(p(y = 1|\mathbf{x}_j, \mathbf{w}, b)) = \sum_{j=l+1}^{l+u} H(1/\left(1 + \exp(-f(\mathbf{x}_j))\right)). \tag{6.29}$$

The entropy regularizer is small if the classification on the unlabeled instances is certain. Figure 6.4(b) shows the entropy regularizer on a single unlabeled instance $\mathbf{x}$ as a function of $f(\mathbf{x})$. Note its similarity to the hat loss in Figure 6.3(b). In direct analogy to S3VMs, we can define *semi-supervised logistic regression* by incorporating this entropy regularizer:

$$\min_{\mathbf{w},b} \sum_{i=1}^{l} \log\left(1 + \exp(-y_i f(\mathbf{x}_i))\right) + \lambda_1 \|\mathbf{w}\|^2 + \lambda_2 \sum_{j=l+1}^{l+u} H(1/\left(1 + \exp(-f(\mathbf{x}_j))\right)). \tag{6.30}$$

6.4 THE ASSUMPTION OF S3VMS AND ENTROPY REGULARIZATION

Remark 6.1. The assumption of both S3VMs and entropy regularization is that the classes are well-separated, such that the decision boundary falls into a low density region in the feature space, and does not cut through dense unlabeled data.

If this assumption does not hold, these algorithms may be led astray. We now describe an example scenario where S3VMs may lead to particularly poor performance.

Example 6.2. S3VMs when the Model Assumption Does Not Hold   Consider the data shown in Figure 6.5. The underlying data distribution $p(x)$ is uniform in a circle of radius 0.5, except for a gap of width 0.2 along the diagonal $y = -x$ where the density is 0. The true class boundary is along the anti-diagonal $y = x$, though. Clearly, the classes are not well-separated, and the low density region does not correspond to the true decision boundary. This poses two problems. First, consider the case in which the labeled instances appear on the same side of the low density region (Figure 6.5(a)). An S3VM's search for a gap between the two classes may stuck in one of many possible local minima. The resulting decision boundary may be worse than the decision boundary of an SVM that does not try to exploit unlabeled data at all. Second and more severely, if the labeled instances appear on opposite sides of the gap (Figure 6.5(b)), the S3VM will be attracted to this region and produce a very poor classifier that gets half of its predictions incorrect.

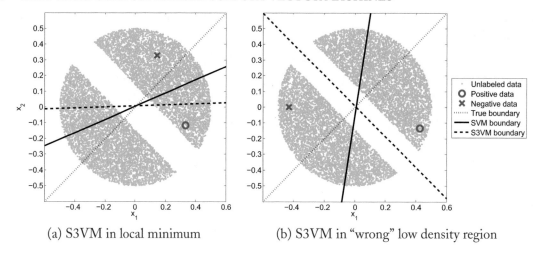

(a) S3VM in local minimum   (b) S3VM in "wrong" low density region

Figure 6.5: (a) With labeled data in adjacent quadrants, S3VM comes close to the true boundary, but it is sensitive to local minima (small gaps between unlabeled data from sampling noise). (b) With labeled data located in opposite quadrants, the S3VM decision boundary seeks a gap in the unlabeled data. In this case, the gap is orthogonal to the classes, so S3VM suffers a large loss.

To obtain a quantitative understanding of the significance of these problems, we sampled 5000 random datasets of this form with $l = 2$ and $u = 10000$. In roughly half the cases, both labeled instances are in the same class, and both a standard SVM and S3VM label everything the same class and get 50% error. In one quarter of the cases, the first problem above occurs, while in the other quarter of the cases, we face the second problem. On average among the half of cases with both classes represented, SVM achieves an unlabeled (transductive) error rate of 0.26 ± 0.13, while S3VM performs much worse with an error rate of 0.34 ± 0.19. (These are mean numbers, plus or minus one standard deviation, over the 5000 random datasets.) The 8% difference in error shows that S3VMs are not well suited to problems of this kind where the classes are not well-separated, and a low-density region does not correspond with the true decision boundary. In fact, despite the large variance in the error rates, S3VM is statistically significantly worse than SVM in this scenario (paired t-test $p \ll 0.05$).

This chapter has introduced the state-of-the-art SVM classifier and its semi-supervised counterpart. Unlike the previous semi-supervised learning techniques we discussed, S3VMs look for a low-density gap in unlabeled data to place the decision boundary. We also introduced entropy regularization, which shares this intuition in a probabilistic framework based on logistic regression. This is the final chapter introducing a new semi-supervised learning approach. In the next chapter, we explore the connections between semi-supervised learning in humans and machines, and discuss the potential impact semi-supervised learning research can have on the cognitive science field.

BIBLIOGRAPHICAL NOTES

Transductive SVMs, as S3VMs were originally called, were proposed by Vapnik [176]. Due to its non-convex nature, early implementations were limited by the problem size they could solve [18, 58, 68]. The first widely used implementation was by Joachims [89]. Since then, various non-convex optimization techniques have been proposed to solve S3VMs [34], including semi-definite programming [55, 54, 188, 189], gradient search with smooth approximation to the hat function [36], deterministic annealing [157], continuation method [32], concave-convex procedure (CCCP) [42], difference convex (DC) programming [182], fast algorithm for linear S3VMs [156], Branch and Bound [33], and stochastic gradient descent which also combines with the manifold assumption [96]. Some recent work relaxes the assumption on unlabeled data [190].

   The idea that unlabeled data should not be very close to the decision boundary is a general one, not limited to S3VMs. It can be implemented in Gaussian Processes with the null category noise model [106, 40], as information regularization [169, 44, 45], maximum entropy discrimination approach [87], or entropy minimization [79, 108, 122].

CHAPTER 7

Human Semi-Supervised Learning

Suppose a young child is learning the names of animals. Dad occasionally points to an animal and says "dog!" But most of the time, the child just watches all sorts of animals by herself. Do such passive experiences help the child learn animals, in addition to the explicit instructions received from Dad? Intuitively, the answer appears to be "yes." Perhaps surprisingly, there is little quantitative study on this question. Clearly, passive experiences are nothing more than unlabeled data, and it seems likely that humans exploit such information in ways similar to how semi-supervised learning algorithms in machines do. In this chapter, we demonstrate the potential value of semi-supervised learning on cognitive science.

7.1 FROM MACHINE LEARNING TO COGNITIVE SCIENCE

Humans are complex learning systems. Cognitive Science, an interdisciplinary science that embraces psychology, artificial intelligence, neuroscience, philosophy, etc., develops theories about human intelligence. Traditionally, cognitive science has benefited from computational models in machine learning, such as reinforcement learning, connectionist models, and non-parametric Bayesian modeling, to name a few. To help understand experiments described in this chapter, we start by providing a "translation" of relevant terms from machine learning to cognitive science:

- Instance $\mathbf{x}$: a stimulus, i.e., an input item to a human subject. For example, a visual stimulus can be a complex shape representing a microscopic pollen particle.

- Class y: a concept category for humans to learn. For example, we may invent two fictitious flowers *Belianthus* and *Nortulaca* that subjects are asked to learn to recognize.

- Classification: a concept learning task for humans. Given a pollen particle as the visual stimulus, the human decides which flower it comes from.

- Labeled data: supervised experience (e.g., explicit instructions) from a teacher. Given a pollen particle $\mathbf{x}$, the teacher says "this is from Belianthus."

- Unlabeled data: passive experiences for humans. The human subject observes a pollen particle $\mathbf{x}$ without receiving its true class.

- Learning algorithm: some mechanism in the mind of the human subject. We cannot directly observe how learning is done.

- Prediction function f: the concepts formed in the human mind. The function itself is not directly observable. However, it is possible to observe any particular prediction $f(\mathbf{x})$, i.e., how the human classifies stimulus $\mathbf{x}$.

In many real world situations, humans are exposed to a combination of labeled data and far more unlabeled data when they need to make a classification decision: an airport security officer must decide whether a piece of luggage poses a threat, where the labeled data comes from job training, and the unlabeled data comes from all the luggage passing through the checkpoint; a dieter must decide which foods are healthy and which are not, based on nutrition labels and advertisements; a child must decide which names apply to which objects from Dad's instructions and observations of the world around her. Some questions naturally arise:

- When learning concepts, do people make systematic use of unlabeled data in addition to labeled data?

- If so, can such usage be understood with reference to semi-supervised learning models in machine learning?

- Can study of the human use of labeled and unlabeled data improve machine learning in domains where human performance outstrips machine performance?

Understanding how humans combine information from labeled and unlabeled data to draw inferences about conceptual boundaries can have significant social impact, ranging from improving education to addressing homeland security issues. Standard psychological theories of concept learning have focused mainly on supervised learning. However, in the realistic setting where labeled and unlabeled data is available, semi-supervised learning offers very explicit computational hypotheses that can be empirically tested in the laboratory. In what follows, we cite three studies that demonstrate the complexity of human semi-supervised learning behaviors.

7.2   STUDY ONE: HUMANS LEARN FROM UNLABELED TEST DATA

Zaki and Nosofsky conducted a behavioral experiment that demonstrates the influence of unlabeled *test* data on learning [194]. In short, unlabeled test data continues to change the concept originally learned in the training phase, in a manner consistent with self-training. For machine learning researchers, it may come as a mild surprise that humans may not be able to hold their learned function fixed during testing.

We present the Zaki and Nosofsky study in machine learning terms. The task is one-class classification or outlier detection: the human subject is first presented with a training sample $\{(\mathbf{x}_i, y_i = 1)\}_{i=1}^l$. Note importantly she is told that all the training instances come from one class. Then, during testing, the subject is shown u unlabeled instances $\{\mathbf{x}_i\}_{i=l+1}^{l+u}$, and her task is to classify each instance as $y_i = 1$ or not. This is usually posed as a density level-set problem in machine learning: From the training sample, the learner estimates the subset of the feature space on which the conditional density $p(\mathbf{x}|y = 1)$ is larger than some threshold ϵ

$$\mathcal{X}_1 = \{\mathbf{x} \in \mathcal{X} \mid p(\mathbf{x}|y = 1) \geq \epsilon\}; \tag{7.1}$$

then during testing, an unlabeled instance $\mathbf{x}$ is classified as $y = 1$ if $\mathbf{x} \in \mathcal{X}_1$, and $y \neq 1$ otherwise.

The approach above assumes that the estimated level-set $\mathcal{X}_1$ is fixed after training. If the assumption were true, then no matter what the test data looks like, the classification for any particular $\mathbf{x}$ would be fixed. Zaki and Nosofsky showed that this is in fact not true. Their experiment compares two conditions that differ only in their test sample distribution $p(\mathbf{x})$; the results demonstrate differences in classification under the two conditions.

In their experiment, each stimulus is a 9-dot pattern as shown in Figure 7.1(a). The location of the nine dots can vary independently, creating the feature space. The training sample consists of 40 instances drawn from a distribution centered around mean μ (which is a particular 9-dot pattern), and with some high variance (large spread). The training density is schematically shown in Figure 7.1(b), and is shared by the two conditions below.

In condition 1, the test sample consists of the following mixture: 4 from the mean itself $\mathbf{x} = \mu$, 20 from a low-variance (small spread) distribution around μ, 20 from the same high-variance distribution around μ, and 40 random instances. This is shown in Figure 7.1(c). Overall, there is a mode (peak of density) at the mean μ. Condition 1 is first used by Knowlton and Squire [100]. Because human experiments are typically noisy, the interesting quantity is the fraction of $y = 1$ classifications for the various groups of instances. In particular, let $\hat{p}(y = 1|\mu)$ be the observed fraction of trials where the subjects classified $y = 1$ among all trials where $\mathbf{x} = \mu$. Let $\hat{p}(y = 1|\text{low})$, $\hat{p}(y = 1|\text{high})$, $\hat{p}(y = 1|\text{random})$ be the similar fraction when $\mathbf{x}$ is drawn from the low-variance, high-variance, and random distribution, respectively. Perhaps not surprisingly, when averaging over a large number of subjects one observes that

$$\hat{p}(y = 1|\mu) > \hat{p}(y = 1|\text{low}) > \hat{p}(y = 1|\text{high})$$
$$\gg \hat{p}(y = 1|\text{random}). \tag{7.2}$$

In condition 2, the mode is intentionally shifted to μ^{new}, which is itself sampled from the high-variance distribution. Specifically, there are 4 instances with $\mathbf{x} = \mu^{\text{new}}$, and 20 instances from a low-variance distribution around μ^{new}. Only 1 test instance (as opposed to 4) remains at the old mean $\mathbf{x} = \mu$, and 2 instances (as opposed to 20) from the low-variance distribution around μ. There are 19 instances (similar to the previous 20) from the high-variance distribution around μ. The 40 random instances remain the same. This is depicted in Figure 7.1(d). Under this test sample distribution, human behaviors are drastically different:

$$\hat{p}(y = 1|\mu^{\text{new}}) > \hat{p}(y = 1|\text{low}^{\text{new}})$$
$$> \quad \hat{p}(y = 1|\mu) \approx \hat{p}(y = 1|\text{low}) \approx \hat{p}(y = 1|\text{high})$$
$$\gg \quad \hat{p}(y = 1|\text{random}), \tag{7.3}$$

where low^{new} is the low-variance distribution around μ^{new}.

Therefore, because of the new mode at μ^{new} in *test data*, human subjects perceive μ^{new} to be more likely in class 1. This experiment clearly demonstrates that humans do not fix their hypothesis after training—unlabeled test data influences humans' learned hypotheses. It is worthy pointing out that the observed behavior cannot be simply explained by a change in the threshold ϵ in (7.1). Instead, it likely involves the change in the conditional density $p(\mathbf{x}|y = 1)$.

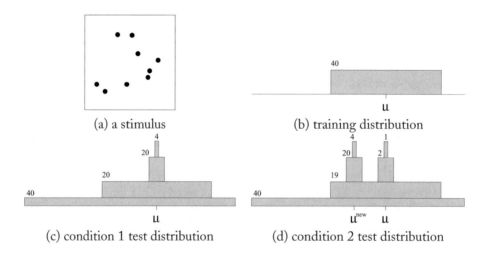

Figure 7.1: The one-class classification problem of Zaki and Nosofsky.

7.3   STUDY TWO: PRESENCE OF HUMAN SEMI-SUPERVISED LEARNING IN A SIMPLE TASK

Zhu, Rogers, Qian and Kalish (ZRQK) conducted an experiment that demonstrates semi-supervised learning behavior in humans [214]. Unlike the Zaki and Nosofsky study, it is a binary classification task. We call it a "simple task" because the feature space is one-dimensional, and there is a single decision boundary (threshold) that separates the two classes. In the next section, we will discuss a more complex task in two dimensions.

Their experiment design is shown in Figure 7.2(a), which is similar to Figure 2.1 earlier in the book. There is a negative labeled instance at $\mathbf{x} = -1$ and a positive labeled instance at $\mathbf{x} = 1$. From these labeled instances alone, supervised learning would put the decision boundary at $\mathbf{x} = 0$. Suppose in addition, the learner observes a large number of unlabeled instances, sampled from the blue bi-modal distribution. Then many semi-supervised learning models will predict that the decision boundary should be in the trough near $\mathbf{x} = -0.5$. Therefore, supervised and semi-supervised learning models lead to different decision boundaries.

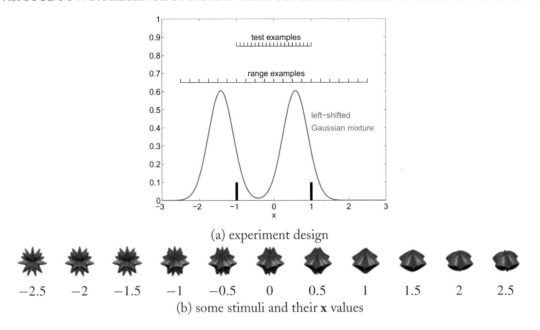

(a) experiment design

| −2.5 | −2 | −1.5 | −1 | −0.5 | 0 | 0.5 | 1 | 1.5 | 2 | 2.5 |

(b) some stimuli and their **x** values

Figure 7.2: The 1D binary classification problem of ZRQK.

Following this intuition, the ZRQK study used the procedure below to study human behaviors. Each stimulus is a complex shape parametrized by a 1-dimensional feature **x**. Figure 7.2(b) shows a few shapes and their x values. Each human subject receives input in the following order:

1. The labeled data: 10 positive labeled instances at **x** = 1 and 10 negative labeled instances at **x** = −1. These 20 trials appear in a different random order for each participant. The repetition of the same two stimuli each ensures quick learning. For these first 20 stimuli, the subjects received an affirmative sound if they made the correct classification, or a warning sound if they were wrong. There was no audio feedback for the remaining stimuli.

2. Test-1: 21 evenly spaced unlabeled examples **x** = −1, −0.9, −0.8, . . . , 1, appearing in a different random order for each participant. They are used to test the learned decision boundary after seeing the above labeled data.

3. The unlabeled data: 690 randomly sampled unlabeled instances from the blue bimodal distribution. Importantly, the two modes are shifted away from the labeled instances at **x** = −1 and **x** = 1, so that the labeled instances are not prototypical examples of the classes. For "L-subjects" the two modes are shifted to the left; for "R-subjects" they are shifted to the right. In addition, three batches of 21 "range instances" evenly spaced in the interval [−2.5, 2.5] are mixed in. The range instances ensure that the unlabeled data for both left-shifted and right-

shifted subjects span the same range, so that any measured shift in the decision boundary cannot be explained by differences in the range of instances viewed.

4. Test-2: same as test-1, to test the learned decision boundary from both the labeled and unlabeled data.

The ZRQK study found that unlabeled data shifts human classification decision boundaries as expected by semi-supervised learning. Figure 7.3(a) shows the logistic function fit to the empirical fraction $\hat{p}(y = 1|\mathbf{x})$, i.e., the fraction of human subjects classifying a given $\mathbf{x}$ as positive. The decision boundary can be estimated as the $\mathbf{x}$ for which the curve intersects $\hat{p}(y = 1|\mathbf{x}) = 0.5$. For all participants in test-1 (the dotted curve), the decision boundary is at $\mathbf{x} = 0.11$, close to the expected boundary at zero from supervised learning. The curve is also relatively steep, showing that the participants are highly consistent in their classifications immediately after seeing the 20 labeled instances. For R-subjects in test-2 (the dashed curve), the decision boundary is at $\mathbf{x} = 0.48$. This represents a shift to the right of 0.37, compared to test-1. This shift represents the effect of unlabeled data on the R-subjects and fits the expectation of semi-supervised learning. For L-subjects in test-2 (the solid curve), the decision boundary is at $\mathbf{x} = -0.10$. This represents a shift to the left by -0.21, also consistent with semi-supervised learning.

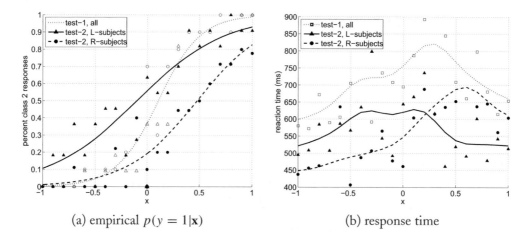

(a) empirical $p(y = 1|\mathbf{x})$ (b) response time

Figure 7.3: In the ZRQK study, unlabeled data changes human concept boundaries, as revealed by classification and response time.

As additional evidence for boundary shift, the ZRQK study observed changes in response time. A long response time implies that the stimulus is relatively difficult to classify. Stimuli near the decision boundary should be associated with longer response times. Figure 7.3(b) shows mean response times on test-1 (dotted line). After seeing just the labeled instances at $\mathbf{x} = -1$ and $\mathbf{x} = 1$, people react quickly to examples near them (600ms), but are much slower (800ms) for instances "in the middle," that is, near the decision boundary. More interesting is the mean response time on test-2 for L-subjects (solid line) and R-subjects (dashed line). L-subjects have a response time plateau around $x = -0.1$, which is left-shifted compared to test-1, whereas R-subjects have a response time peak around $x = 0.6$, which is right-shifted. The response times thus also suggest that exposure to the unlabeled data has shifted the decision boundary.

The ZRQK study explains the human behavior with a Gaussian Mixture Model (see Chapter 3). They fit the model with the EM algorithm. After seeing the labeled data plus test-1 (since these are what the human subjects see when they make decisions on test-1), the Gaussian Mixture Model predicts $p(y = 1|x)$ as the dotted line in Figure 7.4(a). Then, after exposure to unlabeled data, the Gaussian Mixture Models fit on all data (labeled, test-1, unlabeled, and test-2) predicts decision boundary shift for the left-shift condition (solid line) and right-shift condition (dashed line) in Figure 7.4(a), which qualitatively explains the observed human behavior.

These Gaussian Mixture Models can also explain the observed response time. Let the response time model be

$$aH(\mathbf{x}) + b_i, \tag{7.4}$$

for test-i, $i = 1, 2$. $H(\mathbf{x})$ is the entropy of the prediction

$$H(\mathbf{x}) = \sum_{y=-1,1} -p(y|\mathbf{x}) \log p(y|\mathbf{x}), \tag{7.5}$$

which is zero for confident predictions $p(y|\mathbf{x}) = 0$ or 1, and one for uncertain predictions $p(y|\mathbf{x}) = 0.5$. The parameters $a = 168$, $b_1 = 688$, $b_2 = 540$, obtained with least squares from the empirical data, produce the fit in Figure 7.4(b), which explains the empirical peaks before and after seeing unlabeled data in Figure 7.3(b).

7.4 STUDY THREE: ABSENCE OF HUMAN SEMI-SUPERVISED LEARNING IN A COMPLEX TASK

The previous two sections discuss positive studies where human semi-supervised learning behavior is observed. In this section, we present a study by Vandist, De Schryver and Rosseel (VDR), which is a negative result [175]. The task is again binary classification. However, the feature space is two-dimensional. Each class is distributed as a long, thin Gaussian tilted at 45° angle. The true decision boundary is therefore along the diagonal, as shown in Figure 7.5(a). Such non-axis-parallel decision boundaries are called "information-integration tasks" in psychology since the learner has to integrate information from two features [7]. They are considered to be more complex and difficult to learn, because there is no verbal analogue to a univariate rule.

In the VDR study, the stimuli are Gabor patches similar to those in Figure 7.5(b), where the two features are frequency and orientation of the "gratings." We discuss one of their experiments that is particularly relevant to human semi-supervised learning. In the experiment, there are two conditions:

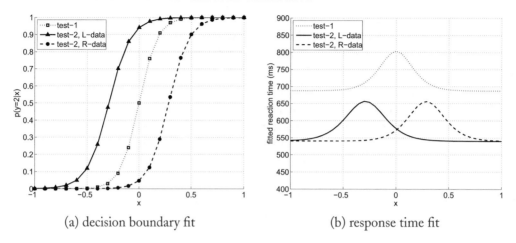

(a) decision boundary fit   (b) response time fit

Figure 7.4: Gaussian Mixture Models make predictions that fit human behaviors in the ZRQK study.

- Condition 1: The subject receives 800 stimuli, one at a time, each randomly sampled from the Gaussian mixture. The subject has to classify each stimulus into one of the two classes. In half of the cases, after making a guess, the subject receives feedback informing her of the true class of the stimulus. In the other half, she receives no feedback. Therefore, there is an equal number of labeled and unlabeled instances.

- Condition 2: Same as condition 1, except that if an instance is unlabeled, its corresponding Gabor patch is replaced by the characters "X" or "N" during display. This turns all the unlabeled data trials into unrelated "filler trials," thus removing the effect of unlabeled data. This condition is therefore similar to pure supervised learning, but properly adjusted to match condition 1.

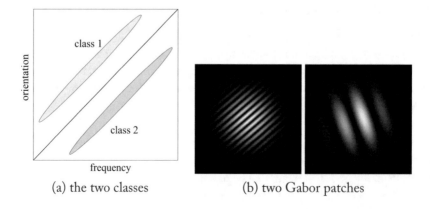

(a) the two classes   (b) two Gabor patches

Figure 7.5: The binary classification task in the VDR study.

If semi-supervised learning helps in this information-integration task, one would expect that subjects under condition 1 learn faster and more accurately than condition 2. This was not observed in the VDR study. The average accuracy in each block of 100 trials is very similar under these two conditions. Both increase gradually, from around 73% in the first block to around 93% in the 8-th (last) block.

Therefore, unlabeled data did not affect learning in this experiment. There might be several factors that contribute to the negative result. For one, the information-integration task in this VDR study is considerably more difficult than the threshold task in the ZRQK study. Second, the VDR study provides much more labeled data too. It may be that the effects of unlabeled data are largest when labels are very sparse.

7.5   DISCUSSIONS

These studies, together with a growing body of recent work, reveal interesting similarities and differences between human learning behavior and semi-supervised learning models. On simple tasks, human's most confident instances depend on test data distribution, and decision boundary judgments align with the low-density region between modes in the unlabeled distribution—just as predicted by machine learning models. They clearly show that supervised learning alone does not account well for human behavior in these tasks, and semi-supervised learning might be a better explanation.

On the harder information-integration task, however, unlabeled data did not help. In addition, in the ZRQK study, human judgments were less certain (i.e., shallower slopes in Figure 7.3(a)) than predicted by machine learning in Figure 7.4(a). These discrepancies may provide leverage for understanding how these models might be adapted to better capture human behavior. For example, we may consider several alternative models: a mixture of heavy-tailed student-t distributions instead of Gaussians; the model's "memory" of previous items could be limited in cognitively-plausible ways; or the model could update its estimates of the mixture coefficients and component parameters sequentially as each new item is presented (e.g., online machine learning).

There are other questions raised by these studies. We may want to design experiments to distinguish the different semi-supervised learning assumptions discussed in this book. Furthermore, what about a child's ability to point to an animal and ask Dad: "What is that?" It seems semi-supervised learning and active learning (i.e., the setting in which the algorithm gets to choose which instances are labeled) go hand-in-hand in human learning. By studying active learning in humans, we may enhance the synergy between semi-supervised and active learning in machines. Ultimately, these studies illustrate the promise of such cross-disciplinary research: congruency between model predictions and human behavior in well-understood learning problems can shed insight into how humans make use of labeled and unlabeled data. In addition, discrepancies between human and machine-predicted behavior points the way toward the development of new machine learning models.

BIBLIOGRAPHICAL NOTES

Advances in machine learning may shed light on the cognitive process behind human learning, which may in turn lead to novel machine learning approaches. Such synergy has long been recognized [130, 105]. The study of human semi-supervised learning is still in its infancy. In the cognitive psychology literature, there have been observations on the effect of unlabeled data to supervised learning, although it was not discussed in a formal semi-supervised learning framework. For examples, see [127, 137, 175, 194]. The earliest quantitative study of human learning with reference to modern semi-supervised machine learning models is [166]. It used drawings of artificial fish to show that human categorization behavior can be influenced by the presence of unlabeled instances. Though certainly suggestive, the experiment had two limitations. First, it used a single positive labeled example and no negative labeled examples, making it a one-class setting similar to novelty detection or quantile estimation, instead of classification. Second, since the stimuli are representations of a familiar concept (i.e., fish), it is difficult to know whether the results reflect prior knowledge about the category, or new learning obtained over the course of the experiment. The first clear demonstration of human semi-supervised learning is [214].

The combination of active learning and semi-supervised learning has attracted attention in machine learning, see, e.g., [85, 126, 132, 183, 203, 202, 213]. In cognitive science, quantitative research on human active learning has just started [31, 103]. We expect the study of the combination to be a fruitful direction.

CHAPTER 8

Theory and Outlook

We have discussed many different semi-supervised learning algorithms throughout the book. One might wonder: is there any theoretic guarantee that these algorithms "work"?

In this chapter, we introduce a simple computational theory to justify semi-supervised learning. Our discussion is based on the notion of *compatibility* proposed by Balcan and Blum [10], and the Probably Approximately Correct (PAC) learning framework of Valiant [174]. The particular theorems in this chapter may not be empirically useful, because they make several strong assumptions that are unlikely to be met in practice. Nonetheless, the way the problem is posed and the proof technique will be useful in understanding other, more advanced versions of the learning theory in the references. In what follows, we will first introduce a simple PAC learning bound for supervised learning, then extend it to semi-supervised learning. For simplicity, we shall restrict ourselves to binary classification tasks.

8.1   A SIMPLE PAC BOUND FOR SUPERVISED LEARNING*

Recall the supervised learning problem. Let the domain of instances be $\mathcal{X}$, and the domain of labels be $\mathcal{Y} = \{-1, 1\}$. Let $P(\mathbf{x}, y)$ be an unknown joint probability distribution on instances and labels $\mathcal{X} \times \mathcal{Y}$. Given a training sample $\mathcal{D}$ consisting of labeled data $\mathcal{D} = \{(\mathbf{x}_i, y_i)\}_{i=1}^{l} \overset{\text{i.i.d.}}{\sim} P(\mathbf{x}, y)$, and a function family $\mathcal{F}$, supervised learning learns a function $f_{\mathcal{D}} \in \mathcal{F}, f : \mathcal{X} \mapsto \mathcal{Y}$. We hope $f_{\mathcal{D}}$ minimizes the true error $e(f)$, which for any function f is defined as:

$$e(f) = \mathbb{E}_{(\mathbf{x}, y) \sim P} \left[f(\mathbf{x}) \neq y \right]. \tag{8.1}$$

But since P is unknown, we cannot directly verify that. Instead, we can observe only the training sample error $\hat{e}(f)$, which is defined in Definition 1.11:

$$\hat{e}(f) = \frac{1}{l} \sum_{i=1}^{l} (f(\mathbf{x}_i) \neq y_i). \tag{8.2}$$

For simplicity, we will let the supervised learning algorithm pick the function $f_{\mathcal{D}} \in \mathcal{F}$ that minimizes the training sample error. In addition, let us assume that the training sample error is zero: $\hat{e}(f_{\mathcal{D}}) = 0$. Can we say anything about its true error $e(f_{\mathcal{D}})$? As mentioned before, the precise value of $e(f_{\mathcal{D}})$ is impossible to compute because we do not know P. However, it turns out that we can *bound* $e(f_{\mathcal{D}})$ without the knowledge of P.

First, notice that $f_{\mathcal{D}}$ is a random variable that depends on the training sample $\mathcal{D}$. Next consider the event $\{e(f_{\mathcal{D}}) > \epsilon\}$, i.e., the particular zero-training-error function $f_{\mathcal{D}}$ chosen by the learning algorithm actually has true error larger than some ϵ. For each random draw of $\mathcal{D} \sim P$, this event $\{e(f_{\mathcal{D}}) > \epsilon\}$ either happens or does not happen. Therefore, we can talk about the probability of this event

$$Pr_{\mathcal{D} \sim P}\left(\{e(f_{\mathcal{D}}) > \epsilon\}\right). \tag{8.3}$$

Our goal will be to show that this probability is small. The learning algorithm picked $f_{\mathcal{D}}$ because $\hat{e}(f_{\mathcal{D}}) = 0$. There could be other functions in $\mathcal{F}$ with zero training error on that $\mathcal{D}$. Consider the union of the events $\cup_{\{f \in \mathcal{F}: \hat{e}(f) = 0\}}\{e(f) > \epsilon\}$. This union contains the event $\{e(f_{\mathcal{D}}) > \epsilon\}$. Therefore,

$$
\begin{aligned}
Pr_{\mathcal{D} \sim P}\left(\{e(f_{\mathcal{D}}) > \epsilon\}\right) &\leq Pr_{\mathcal{D} \sim P}\left(\cup_{\{f \in \mathcal{F}: \hat{e}(f) = 0\}}\{e(f) > \epsilon\}\right) \\
&= Pr_{\mathcal{D} \sim P}\left(\cup_{\{f \in \mathcal{F}\}}\{\hat{e}(f) = 0, e(f) > \epsilon\}\right) \\
&= Pr_{\mathcal{D} \sim P}\left(\cup_{\{f \in \mathcal{F}: e(f) > \epsilon\}}\{\hat{e}(f) = 0\}\right),
\end{aligned}
\tag{8.4}
$$

where the second and the third lines are different ways to represent the same union of events. Now by the union bound $Pr(A \cup B) \leq Pr(A) + Pr(B)$,

$$Pr_{\mathcal{D} \sim P}\left(\cup_{\{f \in \mathcal{F}: e(f) > \epsilon\}}\{\hat{e}(f) = 0\}\right) \leq \sum_{\{f \in \mathcal{F}: e(f) > \epsilon\}} Pr_{\mathcal{D} \sim P}\left(\{\hat{e}(f) = 0\}\right). \tag{8.5}$$

The true error $e(f)$ for a given f can be thought of as a biased coin with heads probability $e(f)$. Because $\mathcal{D}$ is drawn from P, the training sample error $\hat{e}(f)$ is the fraction of heads in l independent coin flips. If the heads probability $e(f) > \epsilon$, the probability of producing l tails in a row is bounded by $(1 - \epsilon)^l$, the product of l independent Bernoulli trials. This is precisely the probability of the event $\{\hat{e}(f) = 0\}$. Thus,

$$\sum_{\{f \in \mathcal{F}: e(f) > \epsilon\}} Pr_{\mathcal{D} \sim P}\left(\{\hat{e}(f) = 0\}\right) \leq \sum_{\{f \in \mathcal{F}: e(f) > \epsilon\}} (1 - \epsilon)^l. \tag{8.6}$$

Finally, assuming that the function family is finite in size,

$$\sum_{\{f \in \mathcal{F}: e(f) > \epsilon\}} (1 - \epsilon)^l \leq \sum_{\{f \in \mathcal{F}\}} (1 - \epsilon)^l = |\mathcal{F}|(1 - \epsilon)^l \leq |\mathcal{F}|e^{-\epsilon l}, \tag{8.7}$$

where the last inequality follows from $1 - x \leq e^{-x}$. Therefore, by connecting the preceding four equations, we arrive at the following inequality:

$$Pr_{\mathcal{D} \sim P}\left(\{e(f_{\mathcal{D}}) > \epsilon\}\right) \leq |\mathcal{F}|e^{-\epsilon l}. \tag{8.8}$$

To see its relevance to learning, consider the complement event

$$Pr_{\mathcal{D} \sim P}\left(\{e(f_{\mathcal{D}}) \leq \epsilon\}\right) \geq 1 - |\mathcal{F}|e^{-\epsilon l}. \tag{8.9}$$

This says that probably (i.e., on at least $1 - |\mathcal{F}|e^{-\epsilon l}$ fraction of random draws of the training sample), the function $f_{\mathcal{D}}$, picked by the supervised learning algorithm because $\hat{e}(f_{\mathcal{D}}) = 0$, is approximately correct (i.e., has true error $e(f_D) \leq \epsilon$). This is known as a Probably Approximately Correct (PAC) bound.

As an application of the above bound, we can compute the number of training instances needed to achieve a given performance criterion:

Theorem 8.1. Simple sample complexity for supervised learning. *Assume $\mathcal{F}$ is finite. Given any $\epsilon > 0, \delta > 0$, if we see l training instances where*

$$l = \frac{1}{\epsilon} \left(\log |\mathcal{F}| + \log \frac{1}{\delta} \right) \tag{8.10}$$

then with probability at least $1 - \delta$, all $f \in \mathcal{F}$ with zero training error $\hat{e}(f) = 0$ have $e(f) \leq \epsilon$.

Here, ϵ is a parameter that controls the error of the learned function, and δ is a parameter that controls the confidence of the bound. The proof follows from setting $\delta = |\mathcal{F}|e^{-\epsilon l}$.

The theorem has limited applicability because it assumes a finite function family, which is seldom the case in practice. For example, training a support vector machine can be thought of as choosing among an infinite number of linear functions. Fortunately, there are ways to extend PAC bounds to the infinite case using concepts such as VC-dimension [97, 176] or Rademacher complexity [14, 153]. The theorem is also implicitly limited by the fact that it does not apply if no f has zero training error. Finally, the theorem does not tell us *how* to find f_D, only what property f_D will have if we find one.

8.2 A SIMPLE PAC BOUND FOR SEMI-SUPERVISED LEARNING*

Semi-supervised learning is helpful if, by taking advantage of the unlabeled data, one can use fewer labeled instances than (8.10) to achieve the same (ϵ, δ) performance. The only way this can happen in (8.10) is to make $|\mathcal{F}|$ smaller. This is exactly our strategy: first, we will use unlabeled data to trim $\mathcal{F}$ down to a much smaller function family; second, we will apply a PAC bound to this smaller function family.

To begin, we need to define a notion of *incompatibility* $\Xi(f, \mathbf{x}) : \mathcal{F} \times \mathcal{X} \mapsto [0, 1]$ between a function f and an unlabeled instance $\mathbf{x}$. This function should be small when f agrees with the semi-supervised learning assumption on $\mathbf{x}$, and large otherwise. As an example, consider the "large unlabeled margin" assumption in S3VMs that all unlabeled instances should be at least γ away from the decision boundary. That is, we want $|f(\mathbf{x})| \geq \gamma$. Further assume γ is known. Then we can define the incompatibility function as

$$\Xi_{\text{S3VM}}(f, \mathbf{x}) = \begin{cases} 1, & \text{if } |f(\mathbf{x})| < \gamma \\ 0, & \text{otherwise.} \end{cases} \tag{8.11}$$

Using the incompatibility function, we can define a "true unlabeled data error"

$$e_U(f) = \mathbb{E}_{\mathbf{x} \sim P_X} \left[\Xi(f, \mathbf{x}) \right]. \tag{8.12}$$

Note it only involves the marginal distribution on $\mathcal{X}$. For a function whose decision boundary cuts through unlabeled data, $e_U(f)$ will be large. Similarly, we can define a "sample unlabeled data error"

$$\hat{e}_U(f) = \frac{1}{u} \sum_{i=l+1}^{l+u} \Xi(f, \mathbf{x}_i). \tag{8.13}$$

Now, with an argument very similar to Theorem 8.1, if we see u unlabeled instances where

$$u = \frac{1}{\epsilon} \left(\log |\mathcal{F}| + \log \frac{2}{\delta} \right), \tag{8.14}$$

then with probability at least $1 - \delta/2$, all $f \in \mathcal{F}$ with $\hat{e}_U(f) = 0$ have $e_U(f) \leq \epsilon$. That is, this amount of unlabeled training data allows us to say with confidence that, if we find a function with $\hat{e}_U(f) = 0$, then it must have come from the sub-family

$$\mathcal{F}(\epsilon) \equiv \{f \in \mathcal{F} : e_U(f) \leq \epsilon\}. \tag{8.15}$$

Note $\mathcal{F}(\epsilon)$ may be much smaller than $\mathcal{F}$. Then, restricting our analysis to functions with $\hat{e}_U(f) = 0$ and applying Theorem 8.1 again (this time on labeled data), we can guarantee the following. After seeing l labeled training instances where

$$l = \frac{1}{\epsilon} \left(\log |\mathcal{F}(\epsilon)| + \log \frac{2}{\delta} \right), \tag{8.16}$$

with probability at least $1 - \delta/2$, all $f \in \mathcal{F}(\epsilon)$ with $\hat{e}(f) = 0$ have $e(f) \leq \epsilon$. Putting the previous two results together, we have the following theorem (Theorem 21.5 in [10]):

Theorem 8.2. Simple sample complexity for semi-supervised learning.   *Assume $\mathcal{F}$ is finite. Given any $\epsilon > 0, \delta > 0$, if we see l labeled and u unlabeled training instances where*

$$l = \frac{1}{\epsilon} \left(\log |\mathcal{F}(\epsilon)| + \log \frac{2}{\delta} \right) \quad \text{and} \quad u = \frac{1}{\epsilon} \left(\log |\mathcal{F}| + \log \frac{2}{\delta} \right), \tag{8.17}$$

then with probability at least $1 - \delta$, all $f \in \mathcal{F}$ with zero training error $\hat{e}(f) = 0$ and zero sample unlabeled data error $\hat{e}_U(f) = 0$ have $e(f) \leq \epsilon$.

Some discussion is in order. This theorem may require less labeled data, compared to Theorem 8.1. Therefore, it justifies semi-supervised learning. However, its conditions are stringent: one must be able to find an f where both the labeled and unlabeled training error is zero. Semi-supervised learning algorithms, such as S3VMs, can be viewed as attempting to minimize both the labeled and unlabeled training error at the same time.

The theorem holds for arbitrary incompatibility functions Ξ. For example, we can define an "inverse S3VM" function which prefers to cut through dense unlabeled data:

$$\Xi_{\text{inv}}(f, \mathbf{x}) = 1 - \Xi_{\text{S3VM}}(f, \mathbf{x}). \tag{8.18}$$

This seems like a terrible idea, and goes against all practices in semi-supervised learning. Paradoxically, Ξ_{inv} could entail a small $\mathcal{F}(\epsilon)$, so the required l may be small, suggesting good semi-supervised learning. What is wrong? Such bad incompatibility functions (which encode inappropriate semi-supervised learning assumptions) will make it difficult to achieve $\hat{e}_U(f) = 0$ and $\hat{e}(f) = 0$ at the same time. They are bad in the sense that they make the above theorem inapplicable.

Finally, we point out that there are several generalizations to Theorem 8.2, as well as theoretic frameworks other than the PAC bounds for semi-supervised learning. These more advanced approaches make weaker assumptions than what is presented here. We give some references in the next section.

8.3   FUTURE DIRECTIONS OF SEMI-SUPERVISED LEARNING

We conclude the book with a brief discussion of what is not covered, and an educated guess on where this field might go.

This book is an introduction, not a survey of the field. It does not discuss many recent topics in semi-supervised learning, including:

- constrained clustering, which is unsupervised learning with some supervision. Interested readers should refer to the book [16] for recent developments in that area. Some techniques there have in turn been used in semi-supervised learning [113];

- semi-supervised regression [25, 47, 159, 205];

- learning in structured output spaces, where the labels y are more complex than scalar values (e.g., sequences, graphs, etc.) [2, 5, 26, 104, 170, 173, 215];

- expectation regularization [124], which may have deep connections with the class proportion constraints in [36, 33, 89, 210];

- learning from positive and unlabeled data, when there is no negative labeled data [61, 114, 109];

- self-taught learning [140] and the universum [186], where the unlabeled data may not come from the positive or negative classes;

- model selection with unlabeled data [94, 120, 150], and feature selection [112];

- inferring label sampling mechanisms [146], multi-instance learning [207], multi-task learning [116], and deep learning [141, 185];

- advances in learning theory for semi-supervised learning [4, 9, 46, 63, 143, 161, 162, 164].

For further readings on these and other semi-supervised learning topics, there is a book collection from a machine learning perspective [37], a survey article with up-to-date papers [208], a book written for computational linguists [1], and a technical report [151].

What is next for semi-supervised learning? We wish the following questions will one day be solved: How to efficiently incorporate massive, even unlimited, amounts of unlabeled data? How to ensure that semi-supervised learning always outperforms supervised learning, by automatically selecting the best semi-supervised model assumption and model parameters? How to design new semi-supervised learning assumptions, perhaps with inspiration from cognitive science? Answering these questions requires innovative research effort. We hope you will join the endeavor.

APPENDIX A

Basic Mathematical Reference

This is a "just enough" quick reference. Please consult standard textbooks for details.

PROBABILITY

The probability of a discrete random variable A taking the value a is $P(A = a) \in [0, 1]$. This is sometimes written as $P(a)$ when there is no danger of confusion.

Normalization: $\sum_a P(A = a) = 1$.

Joint probability: $P(A = a, B = b) = P(a, b)$, the two events both happen at the same time.

Marginalization: $P(A = a) = \sum_b P(A = a, B = b)$.

Conditional probability: $P(a|b) = P(a, b)/P(b)$, the probability of a happening given b happened.

The product rule: $P(a, b) = P(a)P(b|a) = P(b)P(a|b)$.

Bayes rule: $P(a|b) = \frac{P(b|a)P(a)}{P(b)}$. In general, we can condition on one or more random variables C: $P(a|b, C) = \frac{P(b|a, C)P(a|C)}{P(b|C)}$. In the special case when θ is the model parameter and $\mathcal{D}$ is the observed data, we have $p(\theta|\mathcal{D}) = \frac{p(\mathcal{D}|\theta)p(\theta)}{p(\mathcal{D})}$, where $p(\theta)$ is called the prior, $p(\mathcal{D}|\theta)$ the likelihood function of θ (it is *not normalized*: $\int p(\mathcal{D}|\theta) \, d\theta \neq 1$), $p(\mathcal{D}) = \int p(\mathcal{D}|\theta)p(\theta) \, d\theta$ the evidence, and $p(\theta|\mathcal{D})$ the posterior.

Independence: The product rule can be simplified as $P(a, b) = P(a)P(b)$, if and only if A and B are independent. Equivalently, under this condition $P(a|b) = P(a)$, $P(b|a) = P(b)$.

A continuous random variable x has a probability density function (pdf) $p(x) \geq 0$. Unlike discrete random variables, it is possible for $p(x) > 1$ because it is a probability density, not a probability mass. The probability mass in interval $[x_1, x_2]$ is $P(x_1 < X < x_2) = \int_{x_1}^{x_2} p(x) \, dx$, which is between $[0, 1]$.

Normalization: $\int_{-\infty}^{\infty} p(x) \, dx = 1$.

Marginalization: $p(x) = \int_{-\infty}^{\infty} p(x, y) \, dy$.

The expectation of a function $f(x)$ under the probability distribution P for a discrete random variable x is $\mathbb{E}_P[f] = \sum_a P(a) f(a)$, and for a continuous random variable is $\mathbb{E}_P[f] = \int_x p(x) f(x) \, dx$. In particular, if $f(x) = x$, the expectation is the mean of the random variable x.

The variance of x is $\mathrm{Var}(x) = \mathbb{E}[(x - \mathbb{E}[x])^2] = \mathbb{E}[x^2] - \mathbb{E}[x]^2$. The standard deviation of x is $std(x) = \sqrt{\mathrm{Var}(x)}$.

The covariance between two random variables x, y is $\mathrm{Cov}(x, y) = \mathbb{E}_{x,y}[(x - \mathbb{E}[x])(y - \mathbb{E}[y])] = \mathbb{E}_{x,y}[xy] - \mathbb{E}[x]\mathbb{E}[y]$.

When $\mathbf{x}, \mathbf{y}$ are D-dimensional vectors, $\mathbb{E}[\mathbf{x}]$ is the mean vector with the i-th entry being $\mathbb{E}[x_i]$. $\mathrm{Cov}(\mathbf{x}, \mathbf{y})$ is the $D \times D$ covariance matrix with the i, j-th entry being $\mathrm{Cov}(x_i, y_j)$.

DISTRIBUTIONS

Uniform distribution with K outcomes (e.g., a fair K-sided die): $P(A = a_i) = 1/K$, $i = 1, \ldots, K$.

Bernoulli distribution on binary variable $x \in \{0, 1\}$ (e.g., a biased coin with head probability μ): $P(x|\mu) = \mu^x (1 - \mu)^{(1-x)}$. The mean is $\mathbb{E}[x] = \mu$, and the variance is $\mathrm{Var}(x) = \mu(1 - \mu)$.

Binomial distribution: the probability of observing m heads in N trials of a μ-biased coin. $P(m|N, \mu) = \binom{N}{m} \mu^m (1 - \mu)^{N-m}$, with $\binom{N}{m} = \frac{N!}{(N-m)!m!}$. $\mathbb{E}[m] = N\mu$, $\mathrm{Var}(m) = N\mu(1 - \mu)$.

Multinomial distribution: for a K-sided die with probability vector $\mu = (\mu_1, \ldots, \mu_K)$, the probability of observing outcome counts $m_1, \ldots, m_K$ in N trials is $P(m_1, \ldots, m_K|\mu, N) = \binom{N}{m_1 \ldots m_K} \prod_{k=1}^{K} \mu_k^{m_k}$.

Gaussian (Normal) distributions
univariate: $p(x|\mu, \sigma^2) = \frac{1}{\sqrt{2\pi}\sigma} \exp\left(-\frac{(x-\mu)^2}{2\sigma^2}\right)$, with mean μ, variance σ^2.
multivariate: $p(\mathbf{x}|\mu, \Sigma) = \frac{1}{(2\pi)^{\frac{D}{2}} |\Sigma|^{\frac{1}{2}}} \exp\left(-\frac{1}{2}(\mathbf{x} - \mu)^\top \Sigma^{-1}(\mathbf{x} - \mu)\right)$, where $\mathbf{x}$ and μ are D-dimensional vectors, and Σ is a $D \times D$ covariance matrix.

LINEAR ALGEBRA

A scalar is a 1×1 matrix, a vector (default column vector) is an $n \times 1$ matrix.

Matrix transpose: $\left(A^\top\right)_{ij} = A_{ji}$. $(A + B)^\top = A^\top + B^\top$.

Matrix multiplication: An $(n \times m)$ matrix A times an $(m \times p)$ matrix B produces an $(n \times p)$ matrix C, with $C_{ij} = \sum_{k=1}^{m} A_{ik} B_{kj}$. $(AB)C = A(BC)$, $A(B + C) = AB + AC$, $(A + B)C = AC + BC$, $(AB)^\top = B^\top A^\top$. Note in general $AB \neq BA$.

The following properties apply to square matrices.

Diagonal matrix: $A_{ij} = 0, \forall i \neq j$. The identity matrix I is diagonal with $I_{ii} = 1$. $AI = IA = A$ for all square A.

Some square matrices have inverses: $AA^{-1} = A^{-1}A = I$. $(AB)^{-1} = B^{-1}A^{-1}$. $(A^\top)^{-1} = (A^{-1})^\top$.

The trace is the sum of diagonal elements (or eigenvalues): $\mathrm{Tr}(A) = \sum_i A_{ii}$.

The determinant $|A|$ is the product of eigenvalues. $|AB| = |A||B|$, $|a| = a$, $|aA| = a^n|A|$, $|A^{-1}| = 1/|A|$. A matrix A is invertible iff $|A| \neq 0$.

If $|A| = 0$ for an $n \times n$ square matrix A, A is said to be singular. This means at least one column is linearly dependent on (i.e., a linear combination of) other columns (same for rows). Once all such linearly dependent columns and rows are removed, A is reduced to a smaller $r \times r$ matrix, and r is called the rank of A.

An $n \times n$ matrix A has n eigenvalues λ_i and eigenvectors (up to scaling) u_i, such that $Au_i = \lambda_i u_i$. In general, the λ's are complex numbers. If A is real and symmetric, λ's are real numbers, and u's are orthogonal. The u's can be scaled to orthonormal, i.e., length one, so that $u_i^\top u_j = I_{ij}$. The spectral decomposition is $A = \sum_i \lambda_i u_i u_i^\top$. For invertible A, $A^{-1} = \sum_i \frac{1}{\lambda_i} u_i u_i^\top$.

A real symmetric matrix A is positive semi-definite if its eigenvalues $\lambda_i \geq 0$. An equivalent condition is $\forall x \in \mathbb{R}^n, x^\top A x \geq 0$. It is strictly positive definite if $\lambda_i > 0$ for all i.

A positive semi-definite matrix has rank r equal to the number of positive eigenvalues. The remaining $n - r$ eigenvalues are zero.

For a vector $x \in \mathbb{R}^n$, we have
0-norm: $\|x\|_0 =$ count of nonzero elements
1-norm: $\|x\|_1 = \sum_{i=1}^{n} |x_i|$
2-norm (the Euclidean norm, the length, or just "the norm"): $\|x\|_2 = \left(\sum_{i=1}^{n} x_i^2\right)^{1/2}$
∞-norm: $\|x\|_\infty = \max_{i=1}^{n} |x_i|$

CALCULUS

The derivative (slope of tangent line) of f at x is $f'(x) = \frac{df}{dx} = \lim_{\delta \to 0} \frac{f(x+\delta)-f(x)}{\delta}$.
The second derivative (curvature) of f at x is $f''(x) = \frac{d^2 f}{dx^2} = \frac{df'}{dx}$. For any constant c,
$c' = 0$, $(cx)' = c$, $(x^a)' = ax^{a-1}$, $(\log x)' = 1/x$, $(e^x)' = e^x$, $(f(x) + g(x))' = f'(x) + g'(x)$,
$(f(x)g(x))' = f'(x)g(x) + f(x)g'(x)$.

The chain rule: $\frac{df(y)}{dx} = \frac{df(y)}{dy} \frac{dy}{dx}$.

The partial derivative of multivariate function $f(x_1, \ldots, x_n)$ w.r.t. x_i is $\frac{\partial f}{\partial x_i} =$
$\lim_{\delta \to 0} \frac{f(x_1 \ldots x_i + \delta \ldots x_n) - f(x_1 \ldots x_i \ldots x_n)}{\delta}$. The gradient at $\mathbf{x} = (x_1, \ldots, x_n)^\top$ is $\nabla f(\mathbf{x}) = \left(\frac{\partial f}{\partial x_1} \cdots \frac{\partial f}{\partial x_n} \right)^\top$.
The gradient is a vector in the same space as $\mathbf{x}$. It points to a "higher ground" in terms of f value.

The second derivatives of a multivariate function form an $n \times n$ Hessian matrix
$\nabla^2 f(\mathbf{x}) = \left(\frac{\partial^2 f}{\partial x_i \partial x_j} \right)_{i,j=1 \ldots n}$.

Sufficient condition for local optimality in unconstrained optimization: Any point $\mathbf{x}$ at which
$\nabla f(x) = 0$ and $\nabla^2 f(x)$ is positive definite is a local minimum.

A function f is convex if $\forall x, y, \ \forall \lambda \in [0, 1], \ f(\lambda x + (1 - \lambda)y) \leq \lambda f(x) + (1 - \lambda) f(y)$.
Common convex functions: $c, cx, (x - c)^n$ if n is an even integer, $|x|, 1/x, e^x$. If the Hessian matrix
exists, it is positive semi-definite.

If f is convex and differentiable, $\nabla f(\mathbf{x}) = 0$ if and only if $\mathbf{x}$ is a global minimum.

APPENDIX B

Semi-Supervised Learning Software

This appendix contains an annotated list of software implementations of semi-supervised learning algorithms available on the Web. The codes are organized by the type of semi-supervised models used. We have tried our best to provide up-to-date author affiliations.

CLUSTER-BASED

Title: Low Density Separation
Authors: Olivier Chapelle (Yahoo! Research), Alexander Zien (Friedrich Miescher Laboratory of the Max Planck Society)
URL: `http://www.kyb.tuebingen.mpg.de/bs/people/chapelle/lds/`
Description: Matlab/C implementation of the low density separation algorithm. This algorithm tries to place the decision boundary in regions of low density, similar to Transductive SVMs.
Related papers: [36]

Title: Semi-Supervised Clustering
Author: Sugato Basu (Google)
URL: `http://www.cs.utexas.edu/users/ml/risc`
Description: Code that performs metric pairwise constrained k-means clustering. Must-link and cannot-link constraints specify requirements for how examples should be placed in clusters.
Related papers: [15, 16]

GRAPH-BASED

Title: Manifold Regularization
Author: Vikas Sindhwani (IBM T.J. Watson Research Center)
URLs: `http://manifold.cs.uchicago.edu/manifold_regularization/software.html`, `http://people.cs.uchicago.edu/~vikass/manifoldregularization.html`
Description: Matlab code that implements manifold regularization and contains several other functions useful for different types of graph-based learning.

Related papers: [17]

Title: Manifold Regularization Demo
Author: Mike Rainey (University of Chicago)
URL: `http://people.cs.uchicago.edu/~mrainey/jlapvis/JLapVis.html`
Description: Graphical demonstration of manifold regularization on toy data sets. Allows users to manipulate graph and regularization parameters to explore the algorithm in detail.
Related papers: [17]

Title: Similarity Graph Demo
Authors: Matthias Hein (Saarland University), Ulrike von Luxburg (Max Planck Institute for Biological Cybernetics)
URL: `http://www.ml.uni-saarland.de/GraphDemo/GraphDemo.html`
Description: Matlab-based graphical user interface for exploring similarity graphs. These graphs can be used for semi-supervised learning, spectral clustering, and other tasks. Also includes several commonly used toy data sets.
Related papers: See references in Chapter 5.

Title: Maximum Variance Unfolding
Author: Kilian Q. Weinberger (Yahoo! Research)
URLs: `http://www.weinbergerweb.net/Downloads/MVU.html`, `http://www.weinbergerweb.net/Downloads/FastMVU.html`
Description: Implements variations of the dimensionality reduction technique known as maximum variance unfolding. This is a graph-based, spectral method that can use unlabeled data in a preprocessing step for classification or other tasks.
Related papers: [148]

Title: SGT^{light} (Spectral Graph Transducer)
Author: Thorsten Joachims (Cornell University)
URL: `http://sgt.joachims.org/`
Description: Implements the spectral graph transducer, which is a transductive learning method based on a combination of minimum cut problems and spectral graph theory.
Related papers: [90]

Title: SemiL
Authors: Te-Ming Huang (INRIX), Vojislav Kecman (University of Auckland)
URL: `http://www.learning-from-data.com/te-ming/semil.htm`
Description: Graph-based semi-supervised learning implementations optimized for large-scale data problems. The code combines and extends the seminal works in graph-based learning.

Related papers: [198, 210, 86]

Title: Harmonic Function
Author: Xiaojin Zhu (University of Wisconsin-Madison)
URL: `http://pages.cs.wisc.edu/~jerryzhu/pub/harmonic_function.m`
Description: Matlab implementation of the harmonic function formulation of graph-based semi-supervised learning.
Related papers: [210]

Title: Active Semi-Supervised Learning
Author: Xiaojin Zhu (University of Wisconsin-Madison)
URL: `http://www.cs.cmu.edu/~zhuxj/pub/semisupervisedcode/active_learning/`
Description: Implementation of semi-supervised learning combined with active learning. In active learning, the algorithm chooses which examples to label in the hopes of reducing the overall amount of data required for learning.
Related papers: [213]

Title: Nonparametric Transforms of Graph Kernels
Author: Xiaojin Zhu (University of Wisconsin-Madison)
URL: `http://pages.cs.wisc.edu/~jerryzhu/pub/nips04.tgz`
Description: Implementation of an approach to building a kernel for semi-supervised learning. A non-parametric kernel is derived from spectral properties of graphs of labeled and unlabeled data. This formulation simplifies the optimization problem to be solved and can scale to large data sets.
Related papers: [211]

S3VMS

Title: SVM^{light}
Author: Thorsten Joachims (Cornell University)
URL: `http://svmlight.joachims.org/`
Description: General purpose support vector machine solver. Performs transductive classification by iteratively refining predictions on unlabeled instances.
Related papers: [88, 89]

Title: SVMlin
Authors: Vikas Sindhwani (IBM T.J. Watson Research Center), S. Sathiya Keerthi (Yahoo! Research)
URL: `http://people.cs.uchicago.edu/~vikass/svmlin.html`

Description: Large-scale linear support vector machine package that can incorporate unlabeled examples using two different techniques for solving the non-convex S3VM problem.
Related papers: [160, 156]

Title: UniverSVM
Author: Fabian Sinz (Max Planck Institute for Biological Cybernetics)
URL: `http://www.kyb.tuebingen.mpg.de/bs/people/fabee/universvm.html`
Description: Another large-scale support vector machine implementation. Performs transductive classification using the Universum technique, trained with the concave-convex procedure (CCCP).
Related papers: [186]

OTHER MODELS

Title: Gaussian Process Learning
Author: Neil Lawrence (University of Manchester)
URL: `http://www.cs.man.ac.uk/~neill/ivmcpp/`
Description: Implementation of semi-supervised learning with Gaussian processes.
Related papers: [106]

APPENDIX C

Symbols

Symbol Usage

δ	confidence parameter in PAC model
ϵ	error parameter in PAC model
λ	weight
Ω	complexity of hypothesis
Σ	covariance matrix of a Gaussian distribution
θ	model parameter
μ	mean of a Gaussian distribution
Ξ	incompatibility between a function and an unlabeled instance
ξ	SVM slack variable
$\mathcal{A}$	an algorithm
b	SVM offset parameter
C	number of classes
c	loss function
$\mathcal{D}$	observed training data
D	feature dimension; degree matrix
d	distance
$\mathcal{F}$	hypothesis space
f	predictor, hypothesis (classification or regression)
$\mathcal{H}$	hidden data
H	entropy
k	number of clusters, nearest neighbors, hypotheses
$\mathcal{L}$	normalized graph Laplacian
L	unnormalized graph Laplacian
l	number of labeled instances
$\mathcal{N}$	Gaussian distribution
n	total number of instances, both labeled and unlabeled
p	probability distribution
q	auxiliary distribution for EM
$\top$	matrix transpose
u	number of unlabeled instances
W, w	graph edge weights
$\mathbf{w}$	SVM parameter vector
$\mathcal{X}$	instance domain

$\mathbf{x}$	instance
$\mathcal{y}$	label domain
y	label

Bibliography

[1] Steven Abney. *Semisupervised Learning for Computational Linguistics*. Chapman & Hall/CRC, 2007.

[2] Y. Altun, D. McAllester, and M. Belkin. Maximum margin semi-supervised learning for structured variables. In *Advances in Neural Information Processing Systems (NIPS) 18*, 2005.

[3] Amazon Mechanical Turk. https://www.mturk.com/.

[4] Massih Amini, Francois Laviolette, and Nicolas Usunier. A transductive bound for the voted classifier with an application to semi-supervised learning. In D. Koller, D. Schuurmans, Y. Bengio, and L. Bottou, editors, *Advances in Neural Information Processing Systems 21*. 2009.

[5] Rie Ando and Tong Zhang. A framework for learning predictive structures from multiple tasks and unlabeled data. *Journal of Machine Learning Research*, 6:1817–1853, 2005.

[6] A. Argyriou. Efficient approximation methods for harmonic semi-supervised learning. Master's thesis, University College London, 2004.

[7] F. G. Ashby, S. Queller, and P. M. Berretty. On the dominance of unidimensional rules in unsupervised categorization. *Perception & Psychophysics*, 61:1178–1199, 1999.

[8] A. Azran. The rendezvous algorithm: multiclass semi-supervised learning with Markov random walks. In Zoubin Ghahramani, editor, *Proceedings of the 24th Annual International Conference on Machine Learning (ICML 2007)*, pages 49–56. Omnipress, 2007. DOI: 10.1145/1273496

[9] Maria-Florina Balcan and Avrim Blum. A PAC-style model for learning from labeled and unlabeled data. In *COLT 2005*, 2005. DOI: 10.1145/1273496

[10] Maria-Florina Balcan and Avrim Blum. An augmented pac model for semi-supervised learning. In O. Chapelle, B. Schölkopf, and A. Zien, editors, *Semi-Supervised Learning*. MIT Press, 2006.

[11] Maria-Florina Balcan, Avrim Blum, Patrick Pakyan Choi, John Lafferty, Brian Pantano, Mugizi Robert Rwebangira, and Xiaojin Zhu. Person identification in webcam images: An application of semi-supervised learning. In *ICML 2005 Workshop on Learning with Partially Classified Training Data*, 2005.

[12] Maria-Florina Balcan, Avrim Blum, and Ke Yang. Co-training and expansion: Towards bridging theory and practice. In Lawrence K. Saul, Yair Weiss, and Léon Bottou, editors, *Advances in Neural Information Processing Systems 17*. MIT Press, Cambridge, MA, 2005.

[13] S. Baluja. Probabilistic modeling for face orientation discrimination: Learning from labeled and unlabeled data. *Neural Information Processing Systems*, 1998.

[14] Peter L. Bartlett and Shahar Mendelson. Rademacher and Gaussian complexities: risk bounds and structural results. *Journal of Machine Learning Research*, 3:463–482, 2002. DOI: 10.1162/153244303321897690

[15] Sugato Basu, Mikhail Bilenko, Arindam Banerjee, and Raymond J. Mooney. Probabilistic semi-supervised clustering with constraints. In O. Chapelle, B. Schölkopf, and A. Zien, editors, *Semi-Supervised Learning*, pages 71–98. MIT Press, 2006.

[16] Sugato Basu, Ian Davidson, and Kiri Wagstaff, editors. *Constrained Clustering: Advances in Algorithms, Theory, and Applications*. Chapman & Hall/CRC Press, 2008.

[17] Mikhail Belkin, Partha Niyogi, and Vikas Sindhwani. Manifold regularization: A geometric framework for learning from labeled and unlabeled examples. *Journal of Machine Learning Research*, 7:2399–2434, November 2006.

[18] Kristin Bennett and Ayhan Demiriz. Semi-supervised support vector machines. *Advances in Neural Information Processing Systems*, 11:368–374, 1999.

[19] Christopher M. Bishop. *Pattern Recognition and Machine Learning*. Springer, 2006.

[20] A. Blum, J. Lafferty, M.R. Rwebangira, and R. Reddy. Semi-supervised learning using randomized mincuts. In *ICML-04, 21st International Conference on Machine Learning*, 2004. DOI: 10.1145/1015330.1015429

[21] Avrim Blum and Shuchi Chawla. Learning from labeled and unlabeled data using graph mincuts. In *Proc. 18th International Conf. on Machine Learning*, 2001.

[22] Avrim Blum and Tom Mitchell. Combining labeled and unlabeled data with co-training. In *COLT: Proceedings of the Workshop on Computational Learning Theory*, 1998.

[23] O. Bousquet, O. Chapelle, and M. Hein. Measure based regularization. In *Advances in Neural Information Processing Systems 16.*, 2004.

[24] Ulf Brefeld, Christoph Büscher, and Tobias Scheffer. Multiview discriminative sequential learning. In *European Conference on Machine Learning (ECML)*, 2005.

[25] Ulf Brefeld, Thomas Gaertner, Tobias Scheffer, and Stefan Wrobel. Efficient co-regularized least squares regression. In *ICML06, 23rd International Conference on Machine Learning*, Pittsburgh, USA, 2006.

[26] Ulf Brefeld and Tobias Scheffer. Semi-supervised learning for structured output variables. In *ICML06, 23rd International Conference on Machine Learning*, Pittsburgh, USA, 2006.

[27] Christopher J.C. Burges and John C. Platt. Semi-supervised learning with conditional harmonic mixing. In Olivier Chapelle, Bernard Schölkopf, and Alexander Zien, editors, *Semi-Supervised Learning*. MIT Press, Cambridge, MA, 2005.

[28] Chris Callison-Burch, David Talbot, and Miles Osborne. Statistical machine translation with word- and sentence-aligned parallel corpora. In *Proceedings of the ACL*, 2004. DOI: 10.3115/1218955.1218978

[29] Miguel A. Carreira-Perpinan and Richard S. Zemel. Proximity graphs for clustering and manifold learning. In Lawrence K. Saul, Yair Weiss, and Léon Bottou, editors, *Advances in Neural Information Processing Systems 17*. MIT Press, Cambridge, MA, 2005.

[30] V. Castelli and T. Cover. The exponential value of labeled samples. *Pattern Recognition Letters*, 16(1):105–111, 1995. DOI: 10.1016/0167-8655(94)00074-D

[31] Rui Castro, Charles Kalish, Robert Nowak, Ruichen Qian, Timothy Rogers, and Xiaojin Zhu. Human active learning. In *Advances in Neural Information Processing Systems (NIPS) 22*. 2008.

[32] Olivier Chapelle, Mingmin Chi, and Alexander Zien. A continuation method for semi-supervised SVMs. In *ICML06, 23rd International Conference on Machine Learning*, Pittsburgh, USA, 2006.

[33] Olivier Chapelle, Vikas Sindhwani, and S. Sathiya Keerthi. Branch and bound for semi-supervised support vector machines. In *Advances in Neural Information Processing Systems (NIPS)*, 2006.

[34] Olivier Chapelle, Vikas Sindhwani, and Sathiya S. Keerthi. Optimization techniques for semi-supervised support vector machines. *Journal of Machine Learning Research*, 9(Feb):203–233, 2008.

[35] Olivier Chapelle, Jason Weston, and Bernhard Schölkopf. Cluster kernels for semi-supervised learning. In *Advances in Neural Information Processing Systems, 15*, volume 15, 2002.

[36] Olivier Chapelle and Alexander Zien. Semi-supervised classification by low density separation. In *Proceedings of the Tenth International Workshop on Artificial Intelligence and Statistics (AISTAT 2005)*, 2005.

[37] Olivier Chapelle, Alexander Zien, and Bernhard Schölkopf, editors. *Semi-supervised learning*. MIT Press, 2006.

[38] Nitesh V. Chawla and Grigoris Karakoulas. Learning from labeled and unlabeled data: An empirical study across techniques and domains. *Journal of Artificial Intelligence Research*, 23:331–366, 2005.

[39] Ke Chen and Shihai Wang. Regularized boost for semi-supervised learning. In J.C. Platt, D. Koller, Y. Singer, and S. Roweis, editors, *Advances in Neural Information Processing Systems 20*, pages 281–288. MIT Press, Cambridge, MA, 2008.

[40] W. Chu and Z. Ghahramani. Gaussian processes for ordinal regression. Technical report, University College London, 2004.

[41] Michael Collins and Yoram Singer. Unsupervised models for named entity classification. In *EMNLP/VLC-99*, 1999.

[42] Ronan Collobert, Jason Weston, and Leon Bottou. Trading convexity for scalability. In *ICML06, 23rd International Conference on Machine Learning*, Pittsburgh, USA, 2006.

[43] A. Corduneanu and T. Jaakkola. Stable mixing of complete and incomplete information. Technical Report AIM-2001-030, MIT AI Memo, 2001.

[44] A. Corduneanu and T. Jaakkola. On information regularization. In *Nineteenth Conference on Uncertainty in Artificial Intelligence (UAI03)*, 2003.

[45] Adrian Corduneanu and Tommi S. Jaakkola. Distributed information regularization on graphs. In Lawrence K. Saul, Yair Weiss, and Léon Bottou, editors, *Advances in Neural Information Processing Systems 17*. MIT Press, Cambridge, MA, 2005.

[46] C. Cortes, M. Mohri, D. Pechyony, and A. Rastogi. Stability of transductive regression algorithms. In Andrew McCallum and Sam Roweis, editors, *Proceedings of the 25th Annual International Conference on Machine Learning (ICML 2008)*, pages 176–183. Omnipress, 2008.

[47] Corinna Cortes and Mehryar Mohri. On transductive regression. In *Advances in Neural Information Processing Systems (NIPS) 19*, 2006.

[48] Fabio Cozman, Ira Cohen, and Marcelo Cirelo. Semi-supervised learning of mixture models. In *ICML-03, 20th International Conference on Machine Learning*, 2003.

[49] Nello Cristianini and John Shawe-Taylor. *An introduction to support vector machines and other kernel-based learning methods*. Cambridge University Press, 2000.

[50] Mark Culp and George Michailidis. An iterative algorithm for extending learners to a semisupervised setting. In *The 2007 Joint Statistical Meetings (JSM)*, 2007.

[51] G. Dai and D. Yeung. Kernel selection for semi-supervised kernel machines. In Zoubin Ghahramani, editor, *Proceedings of the 24th Annual International Conference on Machine Learning (ICML 2007)*, pages 185–192. Omnipress, 2007. DOI: 10.1145/1273496

[52] Rozita Dara, Stefan Kremer, and Deborah Stacey. Clsutering unlabeled data with SOMs improves classification of labeled real-world data. In *Proceedings of the World Congress on Computational Intelligence (WCCI)*, 2002.

[53] Sanjoy Dasgupta, Michael L. Littman, and David McAllester. PAC generalization bounds for co-training. In *Advances in Neural Information Processing Systems (NIPS)*, 2001.

[54] T. De Bie and N. Cristianini. Semi-supervised learning using semi-definite programming. In O. Chapelle, B. Schoëlkopf, and A. Zien, editors, *Semi-supervised learning*. MIT Press, Cambridge-Massachussets, 2006.

[55] Tijl De Bie and Nello Cristianini. Convex methods for transduction. In Sebastian Thrun, Lawrence Saul, and Bernhard Schölkopf, editors, *Advances in Neural Information Processing Systems 16*. MIT Press, Cambridge, MA, 2004.

[56] Virginia R. de Sa. Learning classification with unlabeled data. In *Advances in Neural Information Processing Systems (NIPS)*, 1993.

[57] Olivier Delalleau, Yoshua Bengio, and Nicolas Le Roux. Efficient non-parametric function induction in semi-supervised learning. In *Proceedings of the Tenth International Workshop on Artificial Intelligence and Statistics (AISTAT 2005)*, 2005.

[58] Ayhan Demirez and Kristin Bennett. Optimization approaches to semisupervised learning. In M. Ferris, O. Mangasarian, and J. Pang, editors, *Applications and Algorithms of Complementarity*. Kluwer Academic Publishers, Boston, 2000.

[59] Ayhan Demiriz, Kristin Bennett, and Mark Embrechts. Semi-supervised clustering using genetic algorithms. *Proceedings of Artificial Neural Networks in Engineering*, November 1999.

[60] A. Dempster, N. Laird, and D. Rubin. Maximum likelihood from incomplete data via the EM algorithm. *Journal of the Royal Statistical Society, Series B*, 1977.

[61] Francois Denis, Remi Gilleron, and Marc Tommasi. Text classification from positive and unlabeled examples. In *The 9th International Conference on Information Processing and Management of Uncertainty in Knowledge-Based Systems(IPMU)*, 2002.

[62] Ran El-Yaniv and Leonid Gerzon. Effective transductive learning via objective model selection. *Pattern Recognition Letters*, 26(13):2104–2115, 2005. DOI: 10.1016/j.patrec.2005.03.025

[63] Ran El-Yaniv, Dmitry Pechyony, and Vladimir Vapnik. Large margin vs. large volume in transductive learning. *Machine Learning*, 72(3):173–188, 2008. DOI: 10.1007/s10994-008-5071-9

[64] David Elworthy. Does Baum-Welch re-estimation help taggers? In *Proceedings of the 4th Conference on Applied Natural Language Processing*, 1994. DOI: 10.3115/974358.974371

[65] Jason D.R. Farquhar, David R. Hardoon, Hongying Meng, John Shawe-Taylor, and Sandor Szedmak. Two view learning: SVM-2K, theory and practice. In *Advances in Neural Information Processing Systems (NIPS)*. 2006.

[66] Akinori Fujino, Naonori Ueda, and Kazumi Saito. A hybrid generative/discriminative approach to semi-supervised classifier design. In *AAAI-05, The Twentieth National Conference on Artificial Intelligence*, 2005.

[67] Akinori Fujino, Naonori Ueda, and Kazumi Saito. Semisupervised learning for a hybrid generative/discriminative classifier based on the maximum entropy principle. *IEEE Transaction on Pattern Analysis and Machine Intelligence*, 30(3):424–437, 2008. DOI: 10.1109/TPAMI.2007.70710

[68] Glenn Fung and Olvi Mangasarian. Semi-supervised support vector machines for unlabeled data classification. Technical Report 99-05, Data Mining Institute, University of Wisconsin Madison, October 1999.

[69] Jochen Garcke and Michael Griebel. Semi-supervised learning with sparse grids. In *Proc. of the 22nd ICML Workshop on Learning with Partially Classified Training Data*, Bonn, Germany, August 2005.

[70] Gad Getz, Noam Shental, and Eytan Domany. Semi-supervised learning – a statistical physics approach. In *Proc. of the 22nd ICML Workshop on Learning with Partially Classified Training Data*, Bonn, Germany, August 2005.

[71] J.J. Godfrey, E.C. Holliman, and J. McDaniel. Switchboard: telephone speech corpus for research and development. *Acoustics, Speech, and Signal Processing, 1992. ICASSP-92., 1992 IEEE International Conference on*, 1:517–520 vol.1, March 1992.

[72] Andrew B. Goldberg, Ming Li, and Xiaojin Zhu. Online manifold regularization: A new learning setting and empirical study. In *The European Conference on Machine Learning and Principles and Practice of Knowledge Discovery in Databases (ECML PKDD)*, 2008. DOI: 10.1007/978-3-540-87479-9_44

[73] Andrew B. Goldberg and Xiaojin Zhu. Seeing stars when there aren't many stars: Graph-based semi-supervised learning for sentiment categorization. In *HLT-NAACL 2006 Workshop on Textgraphs: Graph-based Algorithms for Natural Language Processing*, New York, NY, 2006.

[74] Andrew B. Goldberg, Xiaojin Zhu, Aarti Singh, Zhiting Xu, and Robert Nowak. Multi-manifold semi-supervised learning. In *Twelfth International Conference on Artificial Intelligence and Statistics (AISTATS)*, 2009.

[75] Andrew B. Goldberg, Xiaojin Zhu, Aarti Singh, Zhiting Xu, and Robert Nowak. Multi-manifold semi-supervised learning. In *Twelfth International Conference on Artificial Intelligence and Statistics (AISTATS)*, 2009.

[76] Andrew B. Goldberg, Xiaojin Zhu, and Stephen Wright. Dissimilarity in graph-based semi-supervised classification. In *Eleventh International Conference on Artificial Intelligence and Statistics (AISTATS)*, 2007.

[77] Sally Goldman and Yan Zhou. Enhancing supervised learning with unlabeled data. In *Proc. 17th International Conf. on Machine Learning*, pages 327–334. Morgan Kaufmann, San Francisco, CA, 2000.

[78] Leo Grady and Gareth Funka-Lea. Multi-label image segmentation for medical applications based on graph-theoretic electrical potentials. In *ECCV 2004 workshop*, 2004.

[79] Yves Grandvalet and Yoshua Bengio. Semi-supervised learning by entropy minimization. In Lawrence K. Saul, Yair Weiss, and Léon Bottou, editors, *Advances in Neural Information Processing Systems 17*. MIT Press, Cambridge, MA, 2005.

[80] G.R. Haffari and A. Sarkar. Analysis of semi-supervised learning with the Yarowsky algorithm. In *23rd Conference on Uncertainty in Artificial Intelligence (UAI)*, 2007.

[81] T. Hastie, R. Tibshirani, and J. Friedman. *The Elements of Statistical Learning*. Springer, 2001.

[82] Matthias Hein, Jean-Yves Audibert, and Ulrike von Luxburg. Graph Laplacians and their convergence on random neighborhood graphs. *Journal of Machine Learning Research*, 8(Jun):1325–1368, 2007.

[83] Matthias Hein and Markus Maier. Manifold denoising. In *Advances in Neural Information Processing Systems (NIPS) 19*, 2006.

[84] Mark Herbster, Massimiliano Pontil, and Sergio Rojas Galeano. Fast prediction on a tree. In D. Koller, D. Schuurmans, Y. Bengio, and L. Bottou, editors, *Advances in Neural Information Processing Systems 21*. 2009.

[85] Shen-Shyang Ho and Harry Wechsler. Query by transduction. *IEEE Transaction on Pattern Analysis and Machine Intelligence*, 30(9):1557–1571, 2008. DOI: 10.1109/TPAMI.2007.70811

[86] Te-Ming Huang, Vojislav Kecman, and Ivica Kopriva. *Kernel Based Algorithms for Mining Huge Data Sets: Supervised, Semi-supervised, and Unsupervised Learning (Studies in Computational Intelligence)*. Springer-Verlag New York, Inc., Secaucus, NJ, USA, 2006.

[87] T. Jaakkola, M. Meila, and T. Jebara. Maximum entropy discrimination. *Neural Information Processing Systems, 12*, 12, 1999.

[88] Thorsten Joachims. Making large-scale svm learning practical. In B. Schölkopf, C. Burges, and A. Smola, editors, *Advances in Kernel Methods - Support Vector Learning*. MIT Press, 1999.

[89] Thorsten Joachims. Transductive inference for text classification using support vector machines. In *Proc. 16th International Conf. on Machine Learning*, pages 200–209. Morgan Kaufmann, San Francisco, CA, 1999.

[90] Thorsten Joachims. Transductive learning via spectral graph partitioning. In *Proceedings of ICML-03, 20th International Conference on Machine Learning*, 2003.

[91] Rie Johnson and Tong Zhang. On the effectiveness of laplacian normalization for graph semi-supervised learning. *Journal of Machine Learning Research*, 8(Jul):1489–1517, 2007.

[92] Rie Johnson and Tong Zhang. Two-view feature generation model for semi-supervised learning. In *The 24th International Conference on Machine Learning*, 2007.

[93] Rosie Jones. Learning to extract entities from labeled and unlabeled text. Technical Report CMU-LTI-05-191, Carnegie Mellon University, 2005. Doctoral Dissertation.

[94] Matti Kaariainen. Generalization error bounds using unlabeled data. In *COLT 2005*, 2005.

[95] Ashish Kapoor, Yuan Qi, Hyungil Ahn, and Rosalind Picard. Hyperparameter and kernel learning for graph based semi-supervised classification. In *Advances in NIPS*, 2005.

[96] M. Karlen, J. Weston, A. Erkan, and R. Collobert. Large scale manifold transduction. In Andrew McCallum and Sam Roweis, editors, *Proceedings of the 25th Annual International Conference on Machine Learning (ICML 2008)*, pages 448–455. Omnipress, 2008.

[97] Michael J. Kearns and Umesh V. Vazirani. *An Introduction to Computational Learning Theory*. MIT Press, 1994.

[98] C. Kemp, T.L. Griffiths, S. Stromsten, and J.B. Tenenbaum. Semi-supervised learning with trees. In *Advances in Neural Information Processing System 16*, 2003.

[99] Dan Klein. Lagrange multipliers without permanent scarring, 2004. http://www.cs.berkeley.edu/~klein/papers/lagrange-multipliers.pdf.

[100] B. J. Knowlton and L. R. Squire. The learning of categories: Parallel brain systems for item memory and category knowledge. *Science*, 262:1747–1749, 1993. DOI: 10.1126/science.8259522

[101] R. I. Kondor and J. Lafferty. Diffusion kernels on graphs and other discrete input spaces. In *Proc. 19th International Conf. on Machine Learning*, 2002.

[102] Balaji Krishnapuram, David Williams, Ya Xue, Alexander Hartemink, Lawrence Carin, and Mario Figueiredo. On semi-supervised classification. In Lawrence K. Saul, Yair Weiss, and Léon Bottou, editors, *Advances in Neural Information Processing Systems 17*. MIT Press, Cambridge, MA, 2005.

[103] J. K. Kruschke. Bayesian approaches to associative learning: From passive to active learning. *Learning & Behavior*, 36(3):210–226, 2008. DOI: 10.3758/LB.36.3.210

[104] John Lafferty, Xiaojin Zhu, and Yan Liu. Kernel conditional random fields: Representation and clique selection. In *The 21st International Conference on Machine Learning (ICML)*, 2004. DOI: 10.1145/1015330.1015337

[105] Pat Langley. Intelligent behavior in humans and machines. Technical report, Computational Learning Laboratory, CSLI, Stanford University, 2006.

[106] Neil D. Lawrence and Michael I. Jordan. Semi-supervised learning via Gaussian processes. In Lawrence K. Saul, Yair Weiss, and Léon Bottou, editors, *Advances in Neural Information Processing Systems 17*. MIT Press, Cambridge, MA, 2005.

[107] Guy Lebanon. *Riemannian Geometry and Statistical Machine Learning*. PhD thesis, Carnegie Mellon University, 2005. CMU-LTI-05-189.

[108] Chi-Hoon Lee, Shaojun Wang, Feng Jiao, Dale Schuurmans, and Russell Greiner. Learning to model spatial dependency: Semi-supervised discriminative random fields. In *Advances in Neural Information Processing Systems (NIPS) 19*, 2006.

[109] Wee Sun Lee and Bing Liu. Learning with positive and unlabeled examples using weighted logistic regression. In *Proceedings of the Twentieth International Conference on Machine Learning (ICML)*, 2003.

[110] Boaz Leskes. The value of agreement, a new boosting algorithm. In *COLT 2005*, 2005. DOI: 10.1007/b137542

[111] A. Levin, D. Lischinski, and Y. Weiss. Colorization using optimization. In *ACM Transactions on Graphics*, 2004. DOI: 10.1145/1015706.1015780

[112] Yuanqing Li and Cuntai Guan. Joint feature re-extraction and classification using an iterative semi-supervised support vector machine algorithm. *Machine Learning*, 71(1):33–53, 2008. DOI: 10.1007/s10994-007-5039-1

[113] Zhenguo Li, Jianzhuang Liu, and Xiaoou Tang. Pairwise constraint propagation by semidefinite programming for semi-supervised classification. In Andrew McCallum and Sam Roweis, editors, *Proceedings of the 25th Annual International Conference on Machine Learning (ICML 2008)*. Omnipress, 2008.

[114] Bing Liu, Wee Sun Lee, Philip S Yu, and Xiaoli Li. Partially supervised classification of text documents. In *Proceedings of the Nineteenth International Conference on Machine Learning (ICML)*, 2002.

[115] Dong C. Liu and Jorge Nocedal. On the limited memory BFGS method for large scale optimization. *Mathematical Programming*, 45:503–528, 1989. DOI: 10.1007/BF01589116

[116] Qiuhua Liu, Xuejun Liao, and Lawrence Carin. Semi-supervised multitask learning. In J.C. Platt, D. Koller, Y. Singer, and S. Roweis, editors, *Advances in Neural Information Processing Systems 20*, pages 937–944. MIT Press, Cambridge, MA, 2008.

[117] N. Loeff, D. Forsyth, and D. Ramachandran. Manifoldboost: stagewise function approximation for fully-, semi- and un-supervised learning. In Andrew McCallum and Sam Roweis, editors, *Proceedings of the 25th Annual International Conference on Machine Learning (ICML 2008)*, pages 600–607. Omnipress, 2008.

[118] Qing Lu and Lise Getoor. Link-based classification using labeled and unlabeled data. In *ICML 2003 workshop on The Continuum from Labeled to Unlabeled Data in Machine Learning and Data Mining*, 2003.

[119] David J. C. MacKay. *Information Theory, Inference, and Learning Algorithms*. Cambridge University Press, 2003.

[120] Omid Madani, David M. Pennock, and Gary W. Flake. Co-validation: Using model disagreement to validate classification algorithms. In Lawrence K. Saul, Yair Weiss, and Léon Bottou, editors, *Advances in Neural Information Processing Systems 17*. MIT Press, Cambridge, MA, 2005.

[121] Beatriz Maeireizo, Diane Litman, and Rebecca Hwa. Co-training for predicting emotions with spoken dialogue data. In *The Companion Proceedings of the 42nd Annual Meeting of the Association for Computational Linguistics (ACL)*, 2004. DOI: 10.3115/1219044.1219072

[122] Maryam Mahdaviani and Tanzeem Choudhury. Fast and scalable training of semi-supervised CRFs with application to activity recognition. In J.C. Platt, D. Koller, Y. Singer, and S. Roweis, editors, *Advances in Neural Information Processing Systems 20*, pages 977–984. MIT Press, Cambridge, MA, 2008.

[123] Maryam Mahdaviani, Nando de Freitas, Bob Fraser, and Firas Hamze. Fast computational methods for visually guided robots. In *The 2005 International Conference on Robotics and Automation (ICRA)*, 2005.

[124] Gideon S. Mann and Andrew McCallum. Simple, robust, scalable semi-supervised learning via expectation regularization. In *The 24th International Conference on Machine Learning*, 2007. DOI: 10.1145/1273496.1273571

[125] A. McCallum and K. Nigam. A comparison of event models for naive bayes text classification. *AAAI-98 Workshop on Learning for Text Categorization*, 1998.

[126] Andrew K. McCallum and Kamal Nigam. Employing EM in pool-based active learning for text classification. In Jude W. Shavlik, editor, *Proceedings of ICML-98, 15th International Conference on Machine Learning*, pages 350–358, Madison, US, 1998. Morgan Kaufmann Publishers, San Francisco, US.

[127] S. C. McKinley and R. M. Nosofsky. Selective attention and the formation of linear decision boundaries. *Journal of Experimental Psychology: Human Perception & Performance*, 22(2):294–317, 1996. DOI: 10.1037/0096-1523.22.2.294

[128] D. Miller and H. Uyar. A mixture of experts classifier with learning based on both labelled and unlabelled data. In *Advances in NIPS 9*, pages 571–577, 1997.

[129] T. Mitchell. The role of unlabeled data in supervised learning. In *Proceedings of the Sixth International Colloquium on Cognitive Science*, San Sebastian, Spain, 1999.

[130] Tom Mitchell. The discipline of machine learning. Technical Report CMU-ML-06-108, Carnegie Mellon University, 2006.

[131] Tom M. Mitchell. *Machine Learning*. McGraw-Hill, New York, 1997.

[132] Ion Muslea, Steve Minton, and Craig Knoblock. Active + semi-supervised learning = robust multi-view learning. In *Proceedings of ICML-02, 19th International Conference on Machine Learning*, pages 435–442, 2002.

[133] Kamal Nigam. Using unlabeled data to improve text classification. Technical Report CMU-CS-01-126, Carnegie Mellon University, 2001. Doctoral Dissertation.

[134] Kamal Nigam and Rayid Ghani. Analyzing the effectiveness and applicability of co-training. In *Ninth International Conference on Information and Knowledge Management*, pages 86–93, 2000. DOI: 10.1145/354756.354805

[135] Kamal Nigam, Andrew Kachites McCallum, Sebastian Thrun, and Tom Mitchell. Text classification from labeled and unlabeled documents using EM. *Machine Learning*, 39(2/3):103–134, 2000. DOI: 10.1023/A:1007692713085

[136] Zheng-Yu Niu, Dong-Hong Ji, and Chew-Lim Tan. Word sense disambiguation using label propagation based semi-supervised learning. In *Proceedings of the ACL*, 2005. DOI: 10.3115/1219840.1219889

[137] R. M. Nosofsky. Attention, similarity, and the identification-categorization relationship. *Journal of Experimental Psychology: General*, 115(1):39–57, 1986. DOI: 10.1037/0096-3445.115.1.39

[138] Bo Pang and Lillian Lee. A sentimental education: Sentiment analysis using subjectivity summarization based on minimum cuts. In *Proceedings of the Association for Computational Linguistics*, pages 271–278, 2004.

[139] Thanh Phong Pham, Hwee Tou Ng, and Wee Sun Lee. Word sense disambiguation with semi-supervised learning. In *AAAI-05, The Twentieth National Conference on Artificial Intelligence*, 2005.

[140] Rajat Raina, Alexis Battle, Honglak Lee, Benjamin Packer, and Andrew Y. Ng. Self-taught learning: Transfer learning from unlabeled data. In *The 24th International Conference on Machine Learning*, 2007.

[141] M. Ranzato and M. Szummer. Semi-supervised learning of compact document representations with deep networks. In Andrew McCallum and Sam Roweis, editors, *Proceedings of the 25th Annual International Conference on Machine Learning (ICML 2008)*, pages 792–799. Omnipress, 2008.

[142] J. Ratsaby and S. Venkatesh. Learning from a mixture of labeled and unlabeled examples with parametric side information. *Proceedings of the Eighth Annual Conference on Computational Learning Theory*, pages 412–417, 1995. DOI: 10.1145/225298.225348

[143] Philippe Rigollet. Generalization error bounds in semi-supervised classification under the cluster assumption. *Journal of Machine Learning Research*, 8(Jul):1369–1392, 2007.

[144] E. Riloff, J. Wiebe, and T. Wilson. Learning subjective nouns using extraction pattern bootstrapping. In *Proceedings of the Seventh Conference on Natural Language Learning (CoNLL-2003)*, 2003. DOI: 10.3115/1119176.1119180

[145] Charles Rosenberg, Martial Hebert, and Henry Schneiderman. Semi-supervised self-training of object detection models. In *Seventh IEEE Workshop on Applications of Computer Vision*, January 2005. DOI: 10.1109/ACVMOT.2005.107

[146] Saharon Rosset, Ji Zhu, Hui Zou, and Trevor Hastie. A method for inferring label sampling mechanisms in semi-supervised learning. In Lawrence K. Saul, Yair Weiss, and Léon Bottou, editors, *Advances in Neural Information Processing Systems 17*. MIT Press, Cambridge, MA, 2005.

[147] Stuart Russell and Peter Norvig. *Artificial Intelligence: A Modern Approach*. Prentice-Hall, Englewood Cliffs, NJ, second edition, 2003.

[148] L. K. Saul, K. Q. Weinberger, J. H. Ham, F. Sha, and D. D. Lee. Spectral methods for dimensionality reduction. In O. Chapelle B. Schoelkopf and A. Zien, editors, *Semisupervised Learning*. MIT Press, 2006.

[149] B. Schölkopf and A. J. Smola. *Learning with Kernels*. MIT Press, 2002.

[150] Dale Schuurmans and Finnegan Southey. Metric-based methods for adaptive model selection and regularization. *Machine Learning, Special Issue on New Methods for Model Selection and Model Combination*, 48:51–84, 2001. DOI: 10.1023/A:1013947519741

[151] Matthias Seeger. Learning with labeled and unlabeled data. Technical report, University of Edinburgh, 2001.

[152] B. Shahshahani and D. Landgrebe. The effect of unlabeled samples in reducing the small sample size problem and mitigating the Hughes phenomenon. *IEEE Trans. On Geoscience and Remote Sensing*, 32(5):1087–1095, September 1994. DOI: 10.1109/36.312897

[153] John Shawe-Taylor and Nello Cristianini. *Kernel Methods for Pattern Analysis*. Cambridge University Press, 2004.

[154] V. Sindhwani and D. Rosenberg. An rkhs for multi-view learning and manifold co-regularization. In Andrew McCallum and Sam Roweis, editors, *Proceedings of the 25th Annual International Conference on Machine Learning (ICML 2008)*, pages 976–983. Omnipress, 2008.

[155] Vikas Sindhwani, Jianying Hu, and Aleksandra Mojsilovic. Regularized co-clustering with dual supervision. In D. Koller, D. Schuurmans, Y. Bengio, and L. Bottou, editors, *Advances in Neural Information Processing Systems 21*. 2009.

[156] Vikas Sindhwani and S. Sathiya Keerthi. Large scale semisupervised linear SVMs. In *SIGIR 2006*, 2006. DOI: 10.1145/1148170.1148253

[157] Vikas Sindhwani, Sathiya Keerthi, and Olivier Chapelle. Deterministic annealing for semi-supervised kernel machines. In *ICML06, 23rd International Conference on Machine Learning*, Pittsburgh, USA, 2006. DOI: 10.1145/1143844.1143950

[158] Vikas Sindhwani, Partha Niyogi, and Mikhail Belkin. Beyond the point cloud: from transductive to semi-supervised learning. In *ICML05, 22nd International Conference on Machine Learning*, 2005. DOI: 10.1145/1102351.1102455

[159] Vikas Sindhwani, Partha Niyogi, and Mikhail Belkin. A co-regularized approach to semi-supervised learning with multiple views. In *Proc. of the 22nd ICML Workshop on Learning with Multiple Views*, August 2005.

[160] Vikas Sindhwani, Partha Niyogi, Mikhail Belkin, and Sathiya Keerthi. Linear manifold regularization for large scale semi-supervised learning. In *Proc. of the 22nd ICML Workshop on Learning with Partially Classified Training Data*, August 2005.

[161] Aarti Singh, Robert Nowak, and Xiaojin Zhu. Unlabeled data: Now it helps, now it doesn't. In *Advances in Neural Information Processing Systems (NIPS) 22*. 2008.

[162] Kaushik Sinha and Mikhail Belkin. The value of labeled and unlabeled examples when the model is imperfect. In J.C. Platt, D. Koller, Y. Singer, and S. Roweis, editors, *Advances in Neural Information Processing Systems 20*, pages 1361–1368. MIT Press, Cambridge, MA, 2008.

[163] A. Smola and R. Kondor. Kernels and regularization on graphs. In *Conference on Learning Theory, COLT/KW*, 2003.

[164] N. Sokolovska, O. Cappé, and F. Yvon. The asymptotics of semi-supervised learning in discriminative probabilistic models. In Andrew McCallum and Sam Roweis, editors, *Proceedings of the 25th Annual International Conference on Machine Learning (ICML 2008)*, pages 984–991. Omnipress, 2008.

[165] StarDust@Home. http://stardustathome.ssl.berkeley.edu/.

[166] Sean B. Stromsten. *Classification learning from both classified and unclassified examples*. PhD thesis, Stanford University, 2002.

[167] Arthur D. Szlam, Mauro Maggioni, and Ronald R. Coifman. Regularization on graphs with function-adapted diffusion processes. *Journal of Machine Learning Research*, 9(Aug):1711–1739, 2008.

[168] Martin Szummer and Tommi Jaakkola. Partially labeled classification with Markov random walks. In *Advances in Neural Information Processing Systems, 14*, volume 14, 2001.

[169] Martin Szummer and Tommi Jaakkola. Information regularization with partially labeled data. In *Advances in Neural Information Processing Systems, 15*, volume 15, 2002.

[170] Ben Taskar, Carlos Guestrin, and Daphne Koller. Max-margin Markov networks. In *NIPS'03*, 2003.

[171] Wei Tong and Rong Jin. Semi-supervised learning by mixed label propagation. In *Proceedings of the Twenty-Second AAAI Conference on Artificial Intelligence (AAAI)*, 2007.

[172] I. Tsang and J. Kwok. Large-scale sparsified manifold regularization. In *Advances in Neural Information Processing Systems (NIPS) 19*, 2006.

[173] I. Tsochantaridis, T. Joachims, T. Hofmann, and Y. Altun. Large margin methods for structured and interdependent output variables. *Journal of Machine Learning Research*, 6:1453–1484, 2005.

[174] Leslie Valiant. A theory of the learnable. *Communications of the ACM*, 27(11):1134–1142, 1984. DOI: 10.1145/1968.1972

[175] Katleen Vandist, Maarten De Schryver, and Yves Rosseel. Semisupervised category learning: The impact of feedback in learning the information-integration task. *Attention, Perception, & Psychophysics*, 71(2):328–341, 2009. DOI: 10.3758/APP.71.2.328

[176] Vladimir Vapnik. *Statistical Learning Theory*. Wiley-Interscience, 1998.

[177] Luis von Ahn and Laura Dabbish. Labeling images with a computer game. In *CHI '04: Proceedings of the SIGCHI conference on Human factors in computing systems*, pages 319–326, New York, NY, USA, 2004. ACM Press.

[178] U. von Luxburg, M. Belkin, and O. Bousquet. Consistency of spectral clustering. Technical Report TR-134, Max Planck Institute for Biological Cybernetics, 2004. DOI: 10.1214/009053607000000640

[179] Fei Wang and Changshui Zhang. Label propagation through linear neighborhoods. In *ICML06, 23rd International Conference on Machine Learning*, Pittsburgh, USA, 2006. DOI: 10.1145/1143844.1143968

[180] H. Wang, S. Yan, T. Huang, J. Liu, and X. Tang. Transductive regression piloted by inter-manifold relations. In Zoubin Ghahramani, editor, *Proceedings of the 24th Annual International Conference on Machine Learning (ICML 2007)*, pages 967–974. Omnipress, 2007. DOI: 10.1145/1273496

[181] J. Wang, T. Jebara, and S. Chang. Graph transduction via alternating minimization. In Andrew McCallum and Sam Roweis, editors, *Proceedings of the 25th Annual International Conference on Machine Learning (ICML 2008)*, pages 1144–1151. Omnipress, 2008.

[182] Junhui Wang and Xiaotong Shen. Large margin semi-supervised learning. *Journal of Machine Learning Reserach*, 8:1867–1891, 2007.

[183] W. Wang and Z. Zhou. On multi-view active learning and the combination with semi-supervised learning. In Andrew McCallum and Sam Roweis, editors, *Proceedings of the 25th Annual International Conference on Machine Learning (ICML 2008)*, pages 1152–1159. Omnipress, 2008.

[184] Larry Wasserman. *All of Statistics: A Concise Course in Statistical Inference (Springer Texts in Statistics)*. Springer, 2004.

[185] J. Weston, F. Ratle, and R. Collobert. Deep learning via semi-supervised embedding. In Andrew McCallum and Sam Roweis, editors, *Proceedings of the 25th Annual International Conference on Machine Learning (ICML 2008)*, pages 1168–1175. Omnipress, 2008.

[186] Jason Weston, Ronan Collobert, Fabian Sinz, Leon Bottou, and Vladimir Vapnik. Inference with the universum. In *ICML06, 23rd International Conference on Machine Learning*, Pittsburgh, USA, 2006. DOI: 10.1145/1143844.1143971

[187] Mingrui Wu and Bernhard Schölkopf. Transductive classification via local learning regularization. In *Eleventh International Conference on Artificial Intelligence and Statistics (AISTATS)*, 2007.

[188] Linli Xu and Dale Schuurmans. Unsupervised and semi-supervised multi-class support vector machines. In *AAAI-05, The Twentieth National Conference on Artificial Intelligence*, 2005.

[189] Zenglin Xu, Rong Jin, Jianke Zhu, Irwin King, and Michael Lyu. Efficient convex relaxation for transductive support vector machine. In J.C. Platt, D. Koller, Y. Singer, and S. Roweis, editors, *Advances in Neural Information Processing Systems 20*, pages 1641–1648. MIT Press, Cambridge, MA, 2008.

[190] Liu Yang, Rong Jin, and Rahul Sukthankar. Semi-supervised learning with weakly-related unlabeled data : Towards better text categorization. In D. Koller, D. Schuurmans, Y. Bengio, and L. Bottou, editors, *Advances in Neural Information Processing Systems 21*. 2009.

[191] David Yarowsky. Unsupervised word sense disambiguation rivaling supervised methods. In *Proceedings of the 33rd Annual Meeting of the Association for Computational Linguistics*, pages 189–196, 1995. DOI: 10.3115/981658.981684

[192] Kai Yu, Shipeng Yu, and Volker Tresp. Blockwise supervised inference on large graphs. In *Proc. of the 22nd ICML Workshop on Learning with Partially Classified Training Data*, Bonn, Germany, August 2005.

[193] Shipeng Yu, Balaji Krishnapuram, Romer Rosales, Harald Steck, and R. Bharat Rao. Bayesian co-training. In J.C. Platt, D. Koller, Y. Singer, and S. Roweis, editors, *Advances in Neural Information Processing Systems 20*, pages 1665–1672. MIT Press, Cambridge, MA, 2008.

[194] S.R. Zaki and R.M. Nosofsky. A high-distortion enhancement effect in the prototype-learning paradigm: Dramatic effects of category learning during test. *Memory & Cognition*, 35(8):2088–2096, 2007.

[195] Tong Zhang and Rie Ando. Analysis of spectral kernel design based semi-supervised learning. In Y. Weiss, B. Schölkopf, and J. Platt, editors, *Advances in Neural Information Processing Systems 18*. MIT Press, Cambridge, MA, 2006.

[196] Xinhua Zhang and Wee Sun Lee. Hyperparameter learning for graph based semi-supervised learning algorithms. In *Advances in Neural Information Processing Systems (NIPS) 19*, 2006.

[197] D. Zhou and C. Burges. Spectral clustering with multiple views. In Zoubin Ghahramani, editor, *Proceedings of the 24th Annual International Conference on Machine Learning (ICML 2007)*, pages 1159–1166. Omnipress, 2007. DOI: 10.1145/1273496

[198] Dengyong Zhou, Olivier Bousquet, Thomas Lal, Jason Weston, and Bernhard Schlkopf. Learning with local and global consistency. In *Advances in Neural Information Processing System 16*, 2004.

[199] Dengyong Zhou, Jiayuan Huang, and Bernhard Schoelkopf. Learning with hypergraphs: Clustering, classification, and embedding. In *Advances in Neural Information Processing Systems (NIPS) 19*, 2006.

[200] Dengyong Zhou, Jiayuan Huang, and Bernhard Schölkopf. Learning from labeled and unlabeled data on a directed graph. In *ICML05, 22nd International Conference on Machine Learning*, Bonn, Germany, 2005.

[201] Yan Zhou and Sally Goldman. Democratic co-learing. In *Proceedings of the 16th IEEE International Conference on Tools with Artificial Intelligence (ICTAI 2004)*, 2004. DOI: 10.1109/ICTAI.2004.48

[202] Z.-H. Zhou, K.-J. Chen, and H.-B. Dai. Enhancing relevance feedback in image retrieval using unlabeled data. *ACM Transactions on Information Systems*, 24(2):219–244, 2006. DOI: 10.1145/1148020.1148023

[203] Z.-H. Zhou, K.-J. Chen, and Y. Jiang. Exploiting unlabeled data in content-based image retrieval. In *Proceedings of ECML-04, 15th European Conference on Machine Learning*, Italy, 2004. DOI: 10.1007/b100702

[204] Z.-H. Zhou, D.-C. Zhan, and Q. Yang. Semi-supervised learning with very few labeled training examples. In *Twenty-Second AAAI Conference on Artificial Intelligence (AAAI-07)*, 2007.

[205] Zhi-Hua Zhou and Ming Li. Semi-supervised regression with co-training. In *International Joint Conference on Artificial Intelligence (IJCAI)*, 2005.

[206] Zhi-Hua Zhou and Ming Li. Tri-training: exploiting unlabeled data using three classifiers. *IEEE Transactions on Knowledge and Data Engineering*, 17(11):1529–1541, 2005. DOI: 10.1109/TKDE.2005.186

[207] Zhi-Hua Zhou and Jun-Ming Xu. On the relation between multi-instance learning and semi-supervised learning. In *The 24th International Conference on Machine Learning*, 2007. DOI: 10.1145/1273496.1273643

[208] Xiaojin Zhu. Semi-supervised learning literature survey. Technical Report 1530, Department of Computer Sciences, University of Wisconsin, Madison, 2005.

[209] Xiaojin Zhu and Zoubin Ghahramani. Towards semi-supervised classification with Markov random fields. Technical Report CMU-CALD-02-106, Carnegie Mellon University, 2002.

[210] Xiaojin Zhu, Zoubin Ghahramani, and John Lafferty. Semi-supervised learning using Gaussian fields and harmonic functions. In *The 20th International Conference on Machine Learning (ICML)*, 2003.

[211] Xiaojin Zhu, Jaz Kandola, Zoubin Ghahramani, and John Lafferty. Nonparametric transforms of graph kernels for semi-supervised learning. In Lawrence K. Saul, Yair Weiss, and Léon Bottou, editors, *Advances in Neural Information Processing Systems (NIPS) 17*. MIT Press, Cambridge, MA, 2005.

[212] Xiaojin Zhu and John Lafferty. Harmonic mixtures: combining mixture models and graph-based methods for inductive and scalable semi-supervised learning. In *The 22nd International Conference on Machine Learning (ICML)*. ACM Press, 2005. DOI: 10.1145/1102351.1102484

[213] Xiaojin Zhu, John Lafferty, and Zoubin Ghahramani. Combining active learning and semi-supervised learning using Gaussian fields and harmonic functions. In *ICML 2003 workshop on The Continuum from Labeled to Unlabeled Data in Machine Learning and Data Mining*, 2003.

[214] Xiaojin Zhu, Timothy Rogers, Ruichen Qian, and Chuck Kalish. Humans perform semi-supervised classification too. In *Twenty-Second AAAI Conference on Artificial Intelligence (AAAI-07)*, 2007.

[215] A. Zien, U. Brefeld, and T. Scheffer. Transductive support vector machines for structured variables. In Zoubin Ghahramani, editor, *Proceedings of the 24th Annual International Conference on Machine Learning (ICML 2007)*, pages 1183–1190. Omnipress, 2007. DOI: 10.1145/1273496

Biography

XIAOJIN ZHU

Xiaojin Zhu is an assistant professor in the Computer Sciences department at the University of Wisconsin-Madison. His research interests include statistical machine learning and its applications in cognitive psychology, natural language processing, and programming languages. Xiaojin received his Ph.D. from the Language Technologies Institute at Carnegie Mellon University in 2005. He worked on Mandarin speech recognition as a research staff member at IBM China Research Laboratory in 1996-1998. He received M.S. and B.S. in computer science from Shanghai Jiao Tong University in 1996 and 1993, respectively. His other interests include astronomy and geology.

ANDREW B. GOLDBERG

Andrew B. Goldberg is a Ph.D. candidate in the Computer Sciences department at the University of Wisconsin-Madison. His research interests lie in statistical machine learning (in particular, semi-supervised learning) and natural language processing. He has served on the program committee for national and international conferences including AAAI, ACL, EMNLP, and NAACL-HLT. Andrew was the recipient of a UW-Madison First-Year Graduate School Fellowship for 2005-2006 and a Yahoo! Key Technical Challenges Grant for 2008-2009. Before his graduate studies, Andrew received a B.A. in computer science from Amherst College, where he graduated magna cum laude with departmental distinction in 2003. He then spent two years writing, editing, and developing teaching materials for introductory computer science and Web programming textbooks at Deitel and Associates. During this time, he contributed to several Deitel books and co-authored the 3rd edition of *Internet & World Wide Web How to Program*. In 2005, Andrew entered graduate school at UW-Madison and, in 2006 received his M.S. in computer science. In his free time, Andrew enjoys live music, cooking, photography, and travel.

Index

Printed in the United States
by Baker & Taylor Publisher Services